Mechanisms and Cams for Automatic Machines

Influence of Dynamic Forces on High-Speed Machine Components

Mechanisms and Cams for Automatic Machines

Influence of Dynamic Forces on High-Speed Machine Components

C. N. NEKLUTIN, P.E.

Fellow, American Society of Mechanical Engineers
Founder and Consultant, Ferguson Machine Company, St. Louis, Missouri
V.P. of Engineering (retired), Universal Match Corp., St. Louis, Missouri

AMERICAN ELSEVIER PUBLISHING COMPANY, INC.
New York, 1969

AMERICAN ELSEVIER PUBLISHING COMPANY, INC.
52 Vanderbilt Avenue
New York, N. Y. 10017

ELSEVIER PUBLISHING COMPANY
Barking, Essex, England

ELSEVIER PUBLISHING COMPANY
335 Jan Van Galenstraat, P.O. Box 211
Amsterdam, The Netherlands

Standard Book Number 44-00058-5
Library of Congress Card Number 75-81255

Printed in the United States of America

Preface

The book is intended primarily for designers (with or without engineering degrees) of automatic machines.

The material for the book has been created or collected during more than 35 years of designing or guiding the design of automatic machines. The author entered this field in 1926.

Usually designers are interested in quick practical answers to problems—diagrams, charts, and tables will serve this need; designers may use simple equations for calculation of dynamic forces, the importance of which grows with increases of speed of machines and of the weight of moving parts.

Some designers and students of mechanical engineering courses may be interested not only in final results, but also in methods used to develop them. Therefore, some readers may apply these methods to analyses of mechanisms not covered in this book, but which they may meet in practice.

The most important contribution is the description of the system of designing cams, based on the control of dynamic forces. It was originated and developed by the author for solution of numerous cam problems met in practice in different industries.

The main idea, guiding this gradual development, was to offer solutions which can be made quickly even under conditions of drafting rooms of smaller companies, which do not have easy access to computers and do not employ engineers with a scientific background.

The flexibility of the system permits the application of it to a variety of cam problems. Besides, the described method can be used for the development of acceleration diagrams for special conditions.

The use of elementary calculus in the analysis of problems was necessary, but the aim of the search was to assist in the solution of problems facing machine designers. Therefore, the results of analyses were presented in the form applicable to the design of automatic machines, and, in some cases, helpful in improvements of machine shop practice. A special attention was directed to the study of the influence of dynamic forces on performance of machines, including an appraisal of vibrational stresses caused by them.

Besides, other problems, as the strength of some materials and the selection of dimensions of cams, were also analyzed, as they came up.

In short—the book combines an academic approach with a practical one. The book covers the period up to 1965, but the process of searching for "ideal" movements in automatic machines and for proper dimensioning of parts is not completed—there are possibilities of further improvements or, at least, new solutions for special cases.

The author wishes to express appreciation to the late Frank Marriott, Sr., and to Alfred Hollmann for their valuable help in machine shop improvements and in setting up experiments. Special appreciation is due to Adolph H. Rosenberg, the president of Universal Match Corporation for 25 years, for his support of the theoretical research.

C. N. Neklutin

St. Louis, Missouri
April, 1968

Contents

Chapter III

Chapter IV

Introduction

The aim of every designer of production machinery is to develop an automatic machine which has the highest possible capacity per operator with a minimum maintenance cost and with a minimum waste of items handled by the machine.

It is assumed that the preliminary planning work is completed, which involves:

(a) the selection of the methods of handling materials or parts, based on practical knowledge of their properties;

(b) the planning of the sequence of operations of the machine;

(c) the analysis of different movements to determine the most suitable for each operation;

(d) the consideration of a general layout of the machine, including the comparison of advantages and disadvantages of a single or a multiple machine.

The next phase, prior to the actual design of details, involves the study of forces acting in the machine. Basically, these forces can be divided into three groups:

(1) Useful or working forces needed to produce changes in the materials or parts handled by the machine. Actually, machines are built to provide these forces; in some cases the inertia of moving machine parts may be used to provide part of them. It is impossible to list or to classify these forces due to the variety of operations performed by automatic machines.

(2) Frictional forces acting on machine parts and also on items handled by the machine; usually they are estimated; they do not cause difficulties for the designer except when they heat up machine parts or cause an excessive wear. At high speed, they may be beneficial to a certain extent; they always act in the direction opposite to the direction of movements, therefore, they may help to dampen vibrations.

(3) Dynamic or inertial forces; they are created by the necessity to move machine elements and material or parts handled by the

machine. During each stroke, they change in magnitude and in the direction of their action: when an item is accelerated (first part of a stroke), the resisting force varies between zero and a certain maximum and acts in the direction opposite to the direction of the movement; when an item is decelerated (second part of a stroke) the force varies also between zero and maximum, but it acts in the same direction as the movement. It means the machine element, used to produce a stroke, has to provide varying forces—accelerating and decelerating.

At present, it is sufficient to say that dynamic forces are directly proportional to the square of speed or, more correctly, inversely proportional to the square of the time of the stroke. The variation of forces between zero and + and − maximum is the cause of vibrations of moving parts, which create additional stresses and are liable to disturb items handled by the machine, thus causing jams. These are the reasons why the designer should know the magnitude of the dynamic forces and also the character of their changes.

In case of movements created by cams, the designer can select a desirable acceleration diagram and, therefore, can control dynamic forces to a certain extent. He cannot change dynamic properties of standard mechanisms, but, at least, he has to know their characteristics. At high speed or in case of moving heavy masses, the dynamic forces can be the highest of any other forces, so the machine elements should be designed to stand them.

CHAPTER I

DYNAMIC FORCES AND VIBRATIONS

Basic Laws and Terms

Experiments made by Galileo established that every body tends to retain the motion, which it has, or the absence of it—only an outside interference can produce changes in the motion of the body.

The resistance to changes in motion is called "inertia" and its value depends on the "mass" of the body. The mass is a specific property of the body and it is proportional to the weight, which represents the gravitational pull directed to the center of the earth.

Galileo introduced the term momentum, which is the product of the mass and of the velocity; the laws of inertia states that the body tends to retain the value of its momentum.

On the basis of experiments of Galileo and astronomic observation by Kepler, I. Newton formulated the meaning of the term "force."

Force = rate of change of the momentum of the body. Mathematically, it can be represented as

$$F = \frac{d(mv)}{dt}$$

where m is the measure of the mass;

v is the velocity, and

t the time of the change of the momentum.

So the force is the measure of the outside interference which can change the momentum either way—increasing or decreasing it, depending on the direction of the action of the force.

Since the resistance to changes in motion are termed as inertia, the forces, changing the momentum, are called inertial or dynamic forces; machine designers have to consider these forces, as they become more important in machines operating at higher and higher speed.

Newton's law can be presented in a simpler way in case of lineal motion. The mass can be accepted as a property of the matter with a constant value at velocities encountered in machines; therefore, only the velocity changes in the expression for the momentum.

The change of velocity is termed as "acceleration"; it can be positive, if the velocity increases, and negative, if the velocity decreases; the latter is also termed as "deceleration."

The acceleration $a = \frac{dv}{dt}$ and the Newton's law can be stated as

$$\text{Force } F = \text{mass} \;.\; \text{acceleration} = m \;.\; a \qquad (1)$$

The weight of the body is the gravity pull or force acting on the earth and the acceleration produced by this force is established experimentally, therefore, the mass of the body can be established.

$$\text{mass } m = \frac{\text{Force}}{\text{Acceleration}} = \frac{\text{Weight}}{\text{Gravity Acceleration}} = \frac{W^{lb}}{g}$$

Then $\quad F = \frac{W}{g} \;.\; a$

When mass = 1, $\quad F = a$

So, the acceleration diagram is at the same time the diagram of force when the mass m = 1.

In case of rotary motion, the movement (the displacement), the velocity and the acceleration are expressed in terms of radians, namely

displacement in radians,
velocity in radians per sec., and
acceleration in radians per sec. per sec.

They should be expressed in lineal measurements to fit the equation (1).

A radian is a dimensionless ratio $\frac{\text{length of an arc}}{\text{length of a radius}}$

Imagine that the rotating body can be replaced by an equivalent arrangement which has the mass, equal to the mass of rotating body concentrated in one point on the radius r_g. Then the lineal displacement of the mass is the movement in radians multiplied by this radius—

the length of an arc = length of the radius r_g . displacement in radians;

the velocity will be $v = r_g$. velocity in radians per sec., or $r_g \;.\; w$, where w is the angular velocity;

the acceleration will be $a = r_g$. acceleration in radians per sec. per sec., or $r_g \;.\; \Omega$, where Ω is an angular acceleration.

Then the force $\quad F = m \;.\; r_g \;.\; \Omega$

This force acts on the radius r_g, therefore, the required torque is

$$T = F \cdot r_g = m \cdot r_g^2 \Omega$$

The expression $m \cdot r_g^2$ is the polar moment of inertia of the body

$$I = m \cdot r_g^2$$

I can be found in handbooks for different shapes of rotating machine parts. So, the torque is

$$T = I \cdot \Omega \tag{2}$$

It is strongly recommended that all calculations should be made using a certain system of units. In this book, all examples are presented in the following units.

force (or weight) — in pounds;
displacements — in inches;
time — in seconds.

Using this system of units, the gravity acceleration is g = 386 inches per sec. per sec.

Usually, engineers are accustomed to measuring angles in degrees, but for mathematical treatment of equations, it is necessary to have angles expressed in radians, because degrees represent an artificial measure, while radians tie up lineal dimensions of arcs with lineal dimensions of radii, as a dimensionless ratio.

$$1 \text{ revolution} = 360° = 2\pi \text{ radians} = \frac{\text{circumference}}{\text{radius}}$$

$$\text{an angle } \alpha° \text{ in degrees} = \text{angle } \alpha \text{ in radians} \cdot \frac{180}{\pi}.$$

Usually, the rotational speed is presented in revolutions per minute or RPM, but in equations, the speed should be presented as

$$\text{N in revolution per second or Nrps} = \frac{\text{RPM}}{60}$$

Then the time of one revolution $= \frac{1}{N}$ sec.

Here is a group of dimensions of different items:

torque T in inch-lb.;

lineal displacement s in inches;

lineal velocity v in inch/sec.;

lineal acceleration a in inch/sec.^{2};

angular displacement α or s in radians;

angular velocity w radians/sec. = $2\pi N$;

angular acceleration Ω radians/sec.^{2};

mass is $\dfrac{\text{force}}{\text{acceleration}} = \dfrac{\text{lb . sec.}^2}{\text{inch}}$;

polar moment of inertia I is mass . radius^2 = lb . inch . sec.^{2};

momentum = mass . velocity in $\dfrac{\text{lb . sec.}^2}{\text{inch}} \cdot \dfrac{\text{inch}}{\text{sec.}}$ = lb . sec.;

kinetic energy $\dfrac{mv^2}{2}$ is in $\dfrac{\text{lb . sec.}^2}{\text{inch}} \cdot \dfrac{\text{inch}^2}{\text{sec.}^2}$ = lb . inch

In some cases, the knowledge of dimensions permits a better understanding of phenomena. For instance, if at a certain instant the momentum of a moving mass is known, then it is easy to establish a force necessary to create or dissipate it during a certain time. Assume, the mass = $5 \dfrac{\text{lb . sec.}^2}{\text{inch}}$ and the velocity is 50 inch/sec., so the momentum is 250 lb sec.; further assume that the momentum should be brought to zero value (the part is stopped) in 1/10 sec. by a constant force.

Then the required force is

250 lb . sec. ÷ sec. = 2500 lb.

Diagrams

The designer can find all necessary information from three diagrams which describe all kinematic properties of a mechanism

(1) the variation of the displacement during one stroke or during a cycle;

(2) the variation of the velocity, and

(3) the variation of the acceleration, which is equivalent to the variation of the force needed to produce the movement when the mass m = 1.

All diagrams should be drawn on a base which represents time in seconds or the base may represent unity. In case of mechanisms, it is assumed that the driving shaft rotates uniformly, so the angular displacement of the driver is proportional to the time; it is more convenient for designers to have, as a base of diagrams, the rotation of the driver expressed in degrees, because the values on the diagram are directly tied up to the position of the driving shaft.

There are definite connections between the diagrams and they should be clearly understood.

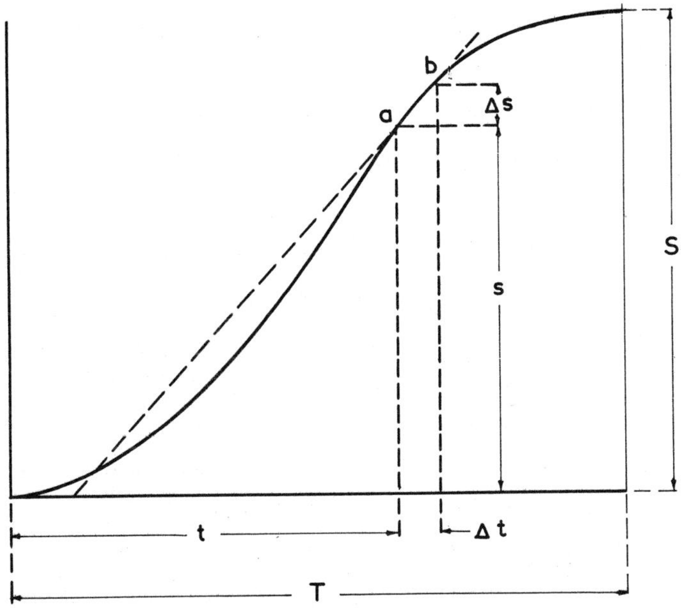

Fig. I - 1

S — total stroke — inches;

T — time of the stroke — sec.

Select a point a which is reached after the time t sec. from the beginning of the stroke;

at that moment the displacement = s.

During the additional, very short period, designated as Δt, the displacement is increased by a small length = Δs.

If Δt is sufficiently small, the line ab can be considered straight; it means an assumption is made that the movement from a to b is at a uniform velocity $= \frac{\Delta s}{\Delta t}$.

This expression also represents a tangent of an angle between the base of the diagram and the extension of the straight line a . b. When the time Δt is decreased, the corresponding point on the diagram approaches the point a; as a limit, the line ab will represent a tangent to the curve at the point a, so the velocity at the point a is written as

$$v = \frac{ds}{dt}$$

where ds and dt are infinitely small.

If the displacement s is expressed by an equation showing the variation of s as a function of time t,

$$s = S \cdot f_s(t),$$

then the velocity at any point can be expressed by, so-called, first derivative

$$\frac{ds}{dt} = \frac{dS[f_s(t)]}{dt} = v$$

The velocity is considered as proportional to a variable expression $f_v(t)$.

It is evident that the magnitude of the angle of the tangent to the displacement curve increases during the first part of the stroke and decrease during the last part of the stroke. So, the velocity diagram should have a maximum at or near the middle of the stroke.

The same reasoning, as applied to the displacement diagram, can be used in case of the velocity diagram. The result will be—tangents to the velocity curve represent the change of the velocity or the acceleration. So the acceleration equation is proportional either to the first derivative of the velocity equation $f_v(t)$ or to the second derivative of the displacement equation. It is evident that the angles of the tangents to the velocity curve increase and decrease during the first part of the stroke. The tangent at the point of maximum velocity is zero and after that, the angles become negative, representing the negative acceleration or deceleration. It means that the moment of the maximum velocity happens when the equation for acceleration is equal zero.

Velocity and acceleration diagrams usually have the following forms:

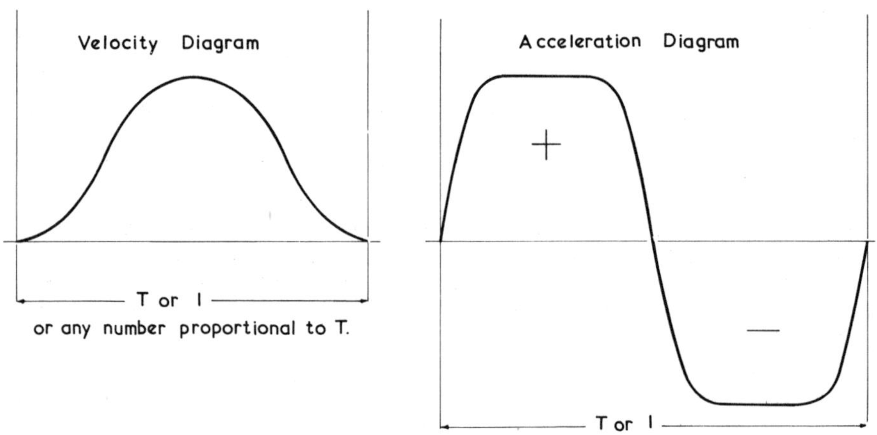

Fig. I - 2

When mechanisms are analyzed, it is possible to write a displacement equation which shows the displacements as functions of the time t. The first and the second derivatives of this equation give equations of the velocity and of the acceleration.

In the case of cams, the process of the analysis is reversed: the first step is to select an acceleration diagram and express it by an equation or by more than one equation. Equations for the velocity and for the displacement should be found from this equation (or equations). If can be done by integration; general rules for that are given in courses of integral calculus and results of the integration in simpler cases are in handbooks.

The physical meaning of the integration can be demonstrated by a step by step conversion of a velocity curve into displacements.

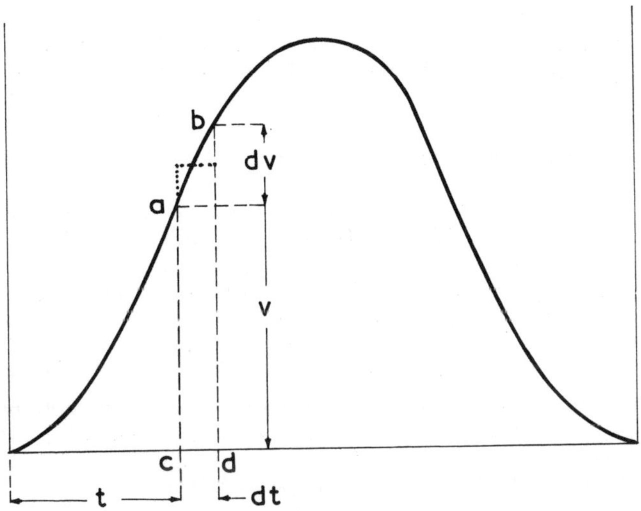

Fig. I - 3

Displacement = velocity . time

The element of time dt is accepted as small, so it is possible to accept that the line ab is straight.

The area of an element cabd is

$$v \ . \ dt + \frac{dv}{2} \ . \ dt$$

As the limit, it can be accepted that dt is infinitely small and, therefore, dv is also infinitely small, so their product dv . dt can be neglected and the area cabd will be presented as v . dt, which represents the dimension of the displacement during the time dt. It is clear, if in the expression for the area cabd the expression for v is introduced; area

$$cabd = \frac{ds}{dt} \ . \ dt = ds.$$

Such small elements, representing these areas under the velocity curve, can be built from the beginning of the velocity curve and each area represents the displacement.

At the moment t the sum of the elementary areas will represent the total displacement up to that moment. The process of summation is solved by the integration rules. In the same way, it can be shown that the area under the acceleration curve at any moment t represents the velocity at that moment.

Selection of a Desirable Acceleration Diagram

Since the magnitude of the acceleration defines the force necessary to produce it, it is desirable to have the maximum acceleration as low as possible. By necessity, the acceleration has to vary between zero and ± maximum, therefore, a question arises about the rate of change of the acceleration. The simplest case which can demonstrate the importance of the rate of change is well known and described in many books. That is a theoretical case of an instantaneous application of a constant force.

Suppose a weight F is suspended above a column; it touches the column, but does not exert any pressure on it. For convenience, assume that the cross-section of the column is 1 sq. inch, so F represents the force in pounds and also stresses in the column in lb per sq. inch.

Suppose, the string supporting the weight is broken instantaneously, so the full force F is applied. It compresses the column on the distance s. At that moment the work done by the weight is F . s inch lb.

This energy is represented on a chart Fig. I-5 as a rectangular area. It is absorbed and equalized by stresses in the column.

Assume that stresses are distributed uniformly in the cross-section. According to Hook's law, the stresses are proportional to the deflection of the column, therefore, at the beginning of the application of the force they are zero and they increase proportionally to the deflection. The energy absorbed by stresses has to be represented by a triangle.

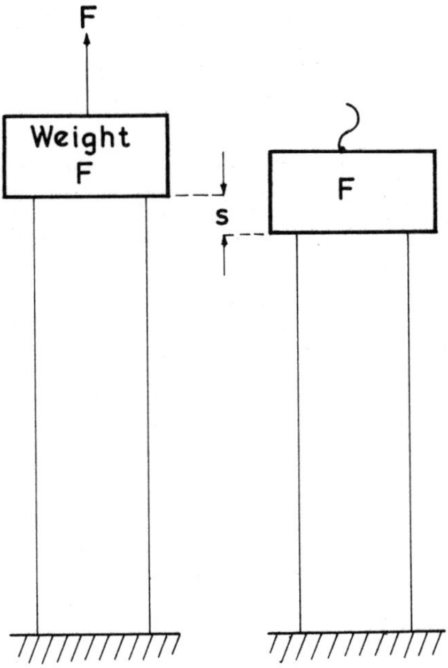

Fig. I - 4

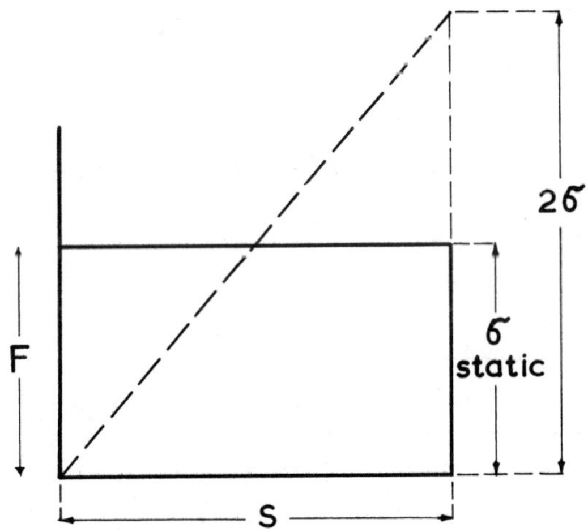

Fig. I - 5

Suppose, when deflections of the column reach s, the stresses will absorb all energy developed by the force F; it means the area of the rectangle should be equal to the area of the triangle. That can happen when stresses develop the opposing force 2F and the maximum stress is twice higher than the stress produced by the force F at rest.

So the energy produced by the force F is equalized, but the forces are not. The greater force will tend to decrease the deflection s and to raise the weight. Vibrations around the deflection $= \frac{s}{2}$ can be expected, until the acting forces can be balanced.

So, the second requirement for a desirable acceleration diagram is the absence of instantaneous variation of the acceleration.

Many diagrams were analyzed, some of them had gradual variation of the acceleration and some had instantaneous changes in the magnitude of the acceleration.

A few discoveries were made during this study. First of all, it was discovered that each diagram of the acceleration has dimensionless coefficients—

of maximum acceleration or deceleration, and
of maximum velocity

They became useful for comparison of different diagrams and for solution of some problems, but also they permitted the development of a method for the presentation of equations for the acceleration, the velocity and for the displacements in convenient forms.

Nomenclature, which will be used during all discussions concerning cams:

- S — stroke, produced by the cam—inches or radians;
- T — total time of the stroke—sec;
- t/T — ratio which varies between 0 and 1;
- $\alpha°$ — working period on the cam (while the stroke S is produced)—degrees;
- N — speed of the cam shaft—revolutions per sec. (RPS);
- C_A & C_D — coefficients of maximum acceleration and of deceleration;
- C_V — coefficient of maximum velocity.

The time T depends on $\alpha°$ and N; the designer mostly deals with them rather than with time in seconds.

The working period represents a part of one revolution $= \frac{\alpha°}{360}$.

The time of one revolution $= \frac{1}{N}$ sec, and

the total time of the stroke $T = \frac{\alpha°}{360N}$ sec. (3)

General Forms of Equations:

Acceleration

$$a = \frac{S}{T^2} \cdot C_A \cdot f_a(t) = S \cdot \left(\frac{360}{\alpha} N\right)^2 \cdot C_A \cdot f_a(t) \qquad (4_1)$$

the ratio $\frac{S}{T^2}$ has dimensions of the acceleration

$$\frac{\text{inch}}{\text{sec.}^2} \text{ or } \frac{\text{radians}}{\text{sec.}^2},$$

therefore, the remaining part of the equation is dimensionless.

$$\frac{a}{S\left(\frac{360}{\alpha} N\right)^2} = C_A \cdot f_a(t) \qquad (4_2)$$

The expression $f_a(t)$ is written in a way that its value varies between 0 and ±1. Therefore, C_A represents the maximum value of the acceleration when

$$\frac{S}{T^2} = 1.$$

The smallest possible value of C_A is for uniform acceleration, namely,

$$C_A = 4.$$

The equation (4_1) shows that the acceleration is directly proportional to the stroke S and to the square of speed N, and it is inversely proportional to the square of the working period α. It is evident that movements producing high acceleration should have the longest possible working period; that gives the designer a chance to balance, as closely as possible, accelerations of different movements in the machine, so the limit for speed of the machine will be established by more than one movement. When the stroke S is in radians (rotary motion), it is advisable to present the acceleration by a letter Ω, instead of a, used for lineal motion.

Velocity

$$v = \frac{S}{T} \cdot C_V \cdot f_V(t) = S \cdot \frac{360}{\alpha} N \cdot C_V \cdot f_V(t) \qquad (5_1)$$

The ratio $\frac{S}{T}$ has dimensions of velocity $\frac{\text{inch}}{\text{sec.}}$ or $\frac{\text{radian}}{\text{sec.}}$, so the remaining part of the equation is dimensionless.

$$\frac{v}{S \cdot \frac{360}{\alpha} \cdot N} = C_v \cdot f_v(t) \qquad (5_2)$$

The variable part of the equation—$f_v(t)$—varies between 0 and +1, so the coefficient C_v represents the maximum value of the speed when $\frac{S}{T} = 1$.

The coefficient C_v has a physical meaning. The ratio $\frac{S}{T}$ is the average (constant) velocity of the whole stroke, so C_v indicates how much higher is the maximum velocity in comparison with the average velocity.

A careful study of different acceleration diagrams permitted to discover the trend of this coefficient: its value is affected by the shape of the first part of the acceleration diagram—it depends on the position of the center of the area of the + acceleration diagram. If this part of the diagram is symmetrical, the center of its area is in the middle of the + acceleration period; then $C_v = 2$. If the center of the area is closer to the beginning of the stroke, then $C_v < 2$. If the center is past the middle of the + acceleration period, then $C_v > 2$.

In a great many cases, the value of C_v is not important, except when a cam moves heavy mass: a lower C_v means lower maximum velocity and a lower kinetic energy of the moving mass $= \frac{m \cdot v_{max}^2}{2}$; so it requires a lower torque on the cam shaft to store less energy. Also, lower C_v means lower maximum slope of the cam, so the cam can be made smaller.

On Fig. I-6, two groups of analyzed diagrams are presented. On the left side are "undesirable" diagrams which have instantaneous variation of the acceleration; after the name of each diagram, the number of these variations is shown in parentheses. On the right side is the group of diagrams without this defect. For each diagram, the values of coefficients C_A and C_v are shown had also an approximate position of the center of the + acceleration area is marked by a heavy dot.

The diagrams are drawn for movements which start from a dwell and end in a dwell, and they represent a case when the + and − acceleration occupy equal periods. It is named Type I; it is the easiest to design and it is used more than other types of movements.

When the stroke S is in radians, the velocity is designated by a letter w, instead of v.

Displacement

$$s = S \cdot f_s(t) \qquad (6_1)$$

or $$\frac{s}{S} = f_s(t) \qquad (6_2)$$

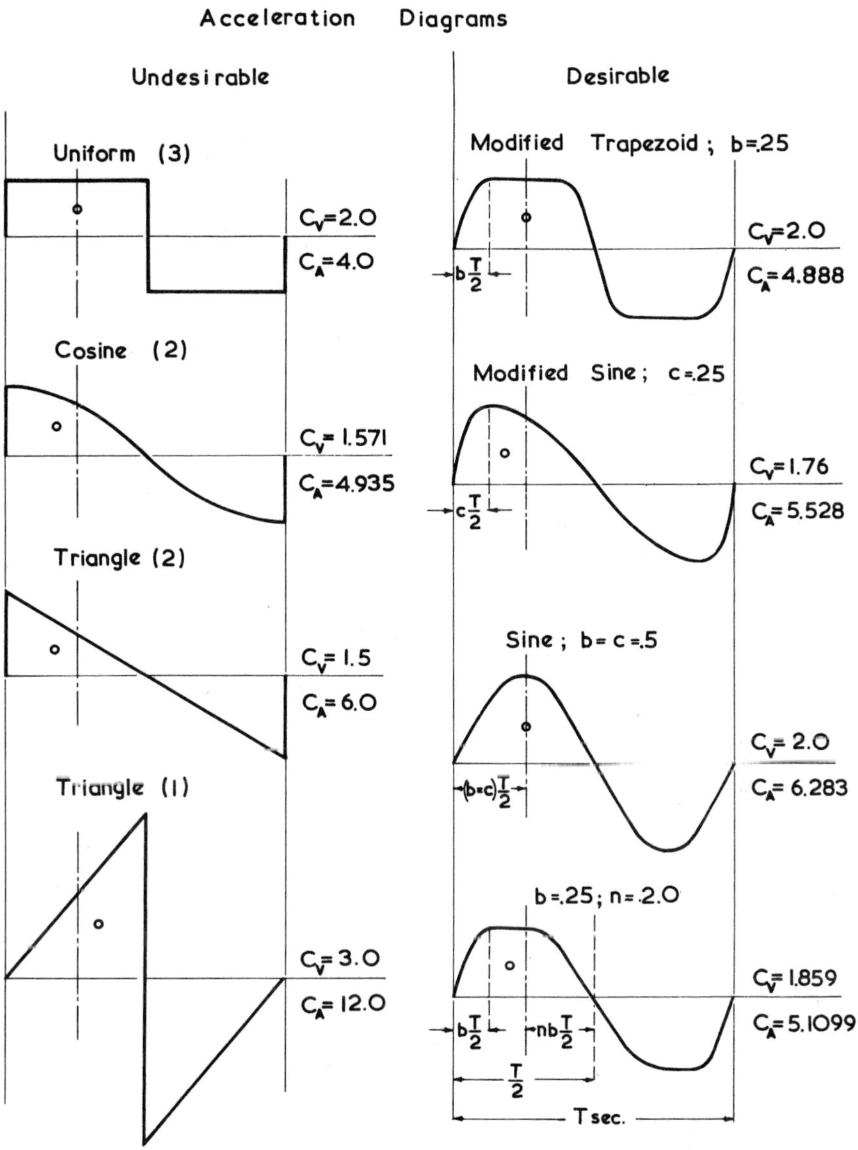
Acceleration Diagrams
Undesirable
Desirable
Uniform (3)
C_V=2.0
C_A=4.0
Cosine (2)
C_V= 1.571
C_A=4.935
Triangle (2)
C_V= 1.5
C_A= 6.0
Triangle (1)
C_V= 3.0
C_A= 12.0
Modified Trapezoid ; b=.25
C_V=2.0
C_A=4.888
Modified Sine; c=.25
C_V=1.76
C_A=5.528
Sine ; b = c =.5
C_V= 2.0
C_A= 6.283
b =.25; n = 2.0
C_V= 1.859
C_A=5.1099
T sec.

Fig. I - 6

The right side of the equation (6_2) is dimensionless, as representing the ratio $\frac{\text{inch}}{\text{inch}}$. The displacement $\frac{s}{S}$ varies between 0 and +1. The knowledge of $\frac{s}{S}$ permits to prepare permanent tables for displacements, which can be used for any stroke, by multiplying figures in the table by S. These tables can be used also for any working period.

The experience indicated that it is sufficient to divide a working period into 360 parts or divisions to have data for making precise master cams. So, the tables can be made for the stroke S = 1 and also for the working period $\alpha = 1$, and they are referred to as "unity tables" for displacement Convenient number of divisions in the "unity" tables should have components 30 or 60, 90, 180, 360, etc.

If the table has 360 divisions and the working period in degrees is $\alpha°$, then the value of one division is

$$\frac{\alpha° \,.\, 60' \,.\, 60''}{360} \text{ seconds of the arc} = (\alpha° \,.\, 10) \text{ seconds of the arc.}$$

So, each division of the table represents 10 seconds of the arc for each degree of the working period $\alpha°$; as a result, the table of displacements, prepared for a machine shop, is convenient for use. For small cams or for small $\alpha°$, it is practical to use only every 2nd, 3rd, or 4th division in the precalculated "unity table."

The knowledge of coefficients C_A and C_V permits to select the desirable acceleration diagram and the existence of unity tables allows the designer to find the position of the cam follower for any critical moment during the stroke, where an interference with other moving machine parts can be expected.

General Remarks

In equations $f_a(t)$, $f_V(t)$ and $f_S(t)$, the variable time t determines the variation of trigonometric functions—sine and cosine. The use of these functions facilitates the writing of equations which vary between zero and unity (± unity in case of $f_a(t)$).

To build different acceleration diagrams, it became necessary to use more than one equation to describe an acceleration diagram. The first practical diagram was developed as a combination of sine curves in combination with periods of uniform acceleration. These periods are included between 1st and 2nd quadrants of the sine function and between 3rd and 4th quadrants. The period occupied by quadrants was selected as a fraction of the acceleration or of the deceleration periods and was designated as b, which can be seen on the Fig. I-6 in the Modified Trapezoid diagram.

During periods b . $\frac{T}{2}$ the acceleration varied either between 0 and + maximum or between 0 and – maximum.

First cams were built with b = 0.15, but they were satisfactory only for speeds not much higher than 200 RPM. The cams with b = 0.25 were introduced and they performed well even at speeds above 1000 RPM, so that value of b was retained.

Next combined diagram was suggested by E. H. Schmidt of DuPont Co.; it uses only sine quadrants; but 1st and 4th quadrants occupy the same fraction of the + or – acceleration period as the Modified Trapezoid, namely 0.25. It was designated as C (see Fig. I-6) and the diagram is named Modified Sine. It was adapted for other movements, besides Type I, and that was the reason for the introduction of the letter C instead of b.

It means that both groups of diagrams—Modified Trapezoid and Sine—always retain a certain relation between different periods of the combined curves. That fact became important for future modifications which will be described in the Chapter III "Cams."

The described forms of equation can be easily used for cams. When mechanisms are analyzed, it is possible to bring equations to the same form, but sometimes only with difficulties, as it will be shown in the Chapter II "Mechanisms."

For a correct appraisal of acceleration diagrams, some knowledge about vibrational stresses produced by dynamic forces is desirable.

Vibrational Stresses

A permanent non-varying force, applied to a structure, produces certain unchanging stresses: they are called static stresses and calculations of them are well developed and described in books dealing with strength of materials.

In automatic machines many different forces may act on machine parts during each stroke and, practically, none of them is constant; due to a great variety of them, it is impossible to have a broad study of all of them. Therefore, only the influence of dynamic forces was studied from the point of view of a machine designer.

To simplify the study effects of possible dampening of vibrations was disregarded; after all the aim of the study was the comparison of different diagrams of acceleration (or of force) rather than the establishing absolute values of vibrational stresses.

Two periods are of interest to the designer:

(1) the period when the force, varying in magnitude and the direction, is applied to a machine element during the stroke;

(2) the dwell period, after the stroke is completed.

1. Vibration Stresses During the Stroke

When a force, applied to a machine part, has varying magnitude, as in the case of dynamic forces, stresses in the part do not change in the same way as the force. At first, stresses lag below the level of the static forces and then they pass them and continue to vary, having values above and below imaginary static stresses, which would exist, if the force changed very slowly. The cycle of these variations of stresses happens during the time of one period of free vibration of the system (or part) to which the force is applied. In other words, the frequency of induced vibrational stresses is the same as the frequency of free vibrations of the part, if it is subjected to one shock, as a hammer blow.

Therefore, the maximum of vibrational stresses depends on the increase of the applied force during one period of free vibrations. The applied force, caused by the necessity to have a certain acceleration, has some maximum value, so a maximum vibrational stress depends on the number of free vibrations during the rise of the force from zero to its maximum.

Designate the maximum stress, due to induced vibration, as x and the static stress produced by the maximum dynamic force as σ_{st}; the latter can be determined because the maximum value of this force can be calculated, if the acceleration diagram of the driving mechanism is known.

The study of the period, when the acting force rises from zero to + maximum, indicates:

> if the affected machine part has one cycle of free vibrations during this period, the vibrational stress is approximately 25% higher than σ_{st}; it is assumed that the acting force changes gradually and not instantaneously.

After this period, the acting force decreases in some way to zero and then increases to a minus maximum—acting in the opposite direction. The maximum vibrational stress may or may not be higher than during the period of the rise of the force; but it is difficult to give any assurance in a general way. Therefore, a practical rule is offered for the safety of the design:

> "the driven part should have at least one and a half free cycles of vibrations during the period when the acting force (or acceleration) changes between zero and its highest maximum value."

If this rule is followed, an additional safety factor = 1.15 would be sufficient in the determination of the strength of the driven part.

2. Vibrational Stresses During a Dwell*

A remark is in order—stresses are proportional to deflections, if they are below the elastic limit of the material; if deflections are large, the vibration of the part can be visible; if they are small, they cannot be seen, but nevertheless, they exist and by themselves or in combination with vibrations of other parts of the machine may dislodge materials or parts, handled by the machine; the result—"unexplainable" jams.

When an acting force disappears at the end of the stroke, the machine part may have some stress left. Instead of decreasing to zero gradually, the stress will vary between + and – value around a no-stress condition. These stresses may be called residue vibrational stresses.

If a dampening device is used, these vibrational stresses may decrease to zero during the dwell; some dampening is caused by hysteresis, so during each cycle the stresses decrease. In high speed machines, dynamic forces may be quite high, but the time of the dwell periods are short, so the stresses left after the termination of the stroke may remain at the end of the dwell.

The stresses created by the force, producing the next stroke, may combine unfavorably with the remaining residue stresses, so vibrational stresses will be higher. That is the reason why residue vibrational stresses were studied.

Many acceleration diagrams were studied including the ones with instantaneous variation of the acceleration. It was found that in almost all cases, the residue stresses are influenced by the number of free vibrations of the driven part during the stroke.

In the case of uniform acceleration, with three instantaneous changes of the magnitude of the acceleration, residue vibrational stresses can be four times higher than σ_{st} when there are 3, 5, 7, etc. free vibrations during the stroke. But when the number of free vibrations during the stroke is even as 4, 6, 8, etc., no residue vibrations exist during the dwell.

In the case of cosine acceleration diagram (see Fig. I-8), the maximum residue vibrations happen when the number of free vibrations during the stroke is a full number as 3, 4, 5, etc. and the stresses are twice higher than σ_{st}. The residue stresses are zero when the number of free vibrations during the stroke were $3\frac{1}{2}$, $4\frac{1}{2}$, $5\frac{1}{2}$, etc.

The triangular acceleration diagram (Fig. I-9) with one instantaneous change of acceleration always has residue stresses twice higher than σ_{st}.

The frequency of free vibrations of a certain machine part is its constant property. So the number of free vibrations during a stroke is affected by the speed of the machine, which may change from day to day, due to

---

* Vibration Analysis of Cams, C. N. Neklutin; Second Conference on Mechanisms at Purdue University, October 1954.

variation of voltage in electric lines or due to some changes in frictional resistances in the machine or due to unsteady transmission ratios between the driving motor and the machine, the designer cannot predict in advance a precise speed of the machine.

It is evident that acceleration diagrams with instantaneous variation of magnitude should not be used, especially at high speed on account of high residue stresses which increase with the speed. The ratio $\frac{x}{\sigma_{st}}$ has the same characteristics, including its maximum, independent of the number of free vibrations during a stroke.

Acceleration diagrams with a gradual variation of the magnitude also produce residue vibrational stresses which are affected by the number of free vibrations of the loaded machine part during one stroke.

Diagrams of the acceleration, changing as Sine functions, have no residue vibration when this number was 2, 3, 4, 5, etc.; residue vibrations reached their maxima when there were $2\frac{1}{2}$, $3\frac{1}{2}$, $4\frac{1}{2}$, $5\frac{1}{2}$, etc. cycles of free vibrations during a stroke.

There is an important property of these residue vibrations: their maxima decrease, if the machine part has higher frequency of free vibrations or, in other words, when the number of free vibrations during the stroke is higher. Therefore, the designer, using this type of acceleration diagrams, can decrease the size of residue stresses by designing lighter parts with a higher rigidity. Such parts have higher frequency of free vibrations, so the residue stresses decrease; besides, due to a greater number of vibrational cycles during a dwell, there is more chance to have them dampened.

Vibrational stresses on Fig. I-7, 8, and 9 are shown during and after an application of the dynamic force, when the acceleration diagrams have an instantaneous change of the magnitude.

On every figure, the maximum value of the disturbing force is represented as unity—for easier comparison of vibrational stresses x and, so-called, static stresses σ_{st}, produced by the maximum magnitude of the applied force. Actually, the maximum value of the force is proportional to the coefficient C_A of maximum acceleration, which is $\geqslant 4$. To compare vibrational stresses created by the different acceleration diagrams, the ratio $\frac{x}{\sigma_{st}}$ should be multiplied by a corresponding coefficient C_A.

The Fig. I-10 shows the comparison of residue vibrational stresses for four diagrams, which have no instantaneous change of the magnitude.

Sine Curve and Isosceles Triangle;

Trapezoid and Modified Trapezoid with $b = 0.25$, meaning that the acceleration changes between 0 and maximum during $b \cdot \frac{T}{2}$ sec., where T is the total time of the stroke.

The comparable values of residue stresses are shown as their maxima which are plotted against the base line representing the number of free vibrations of machine parts during a stroke. This method permits to see clearly the advantage of parts with high frequency of free vibrations.

The results of the analysis of diagrams, which are not recommended for the designing of cams, are shown because the acceleration diagrams of some analyzed mechanisms are of a similar type. For instance, the Fig. I-8 shows the diagram which theoretically represents a Scotch Yoke mechanism, when it is used with a ratchet drive for the intermittent motion. External Geneva drives approach this diagram when they have 6 or more slots in the driven wheel.

The upper diagram on the Fig. I-9 has one instantaneous change of the acceleration from + max. to – max.; External Geneva drives with 3 and 4 slots in the driven wheel and Eccentric gears with a sufficiently flat ellipse also have a very fast change from + to – maximum; that makes them unacceptable for high speed operation.

Fig. I-11 gives the comparison of $C_A \cdot \frac{x}{\sigma_{st}}$ of six acceleration diagrams; lower two of them are presented each with two values of the factor "b" for which the analysis of residue vibrations was made and presented on the Fig. I-10.

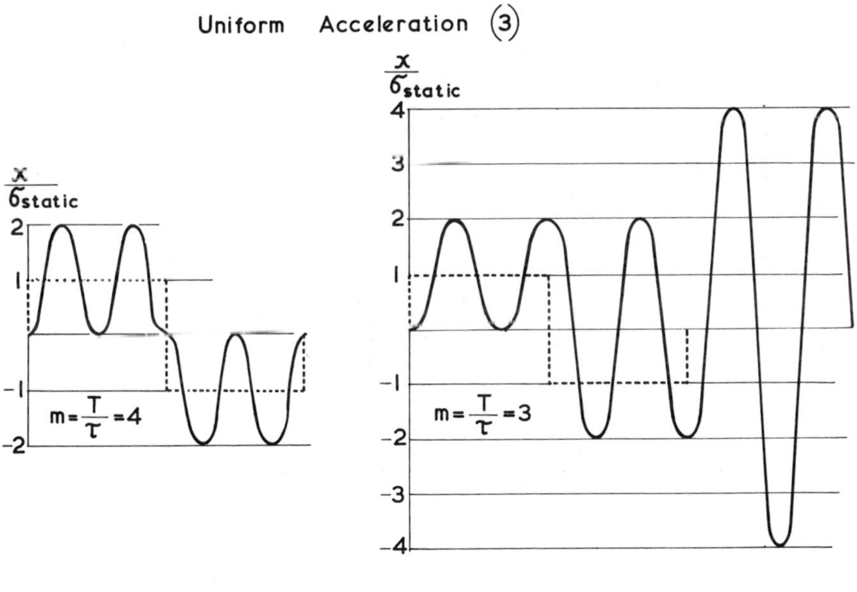

Fig. I - 7

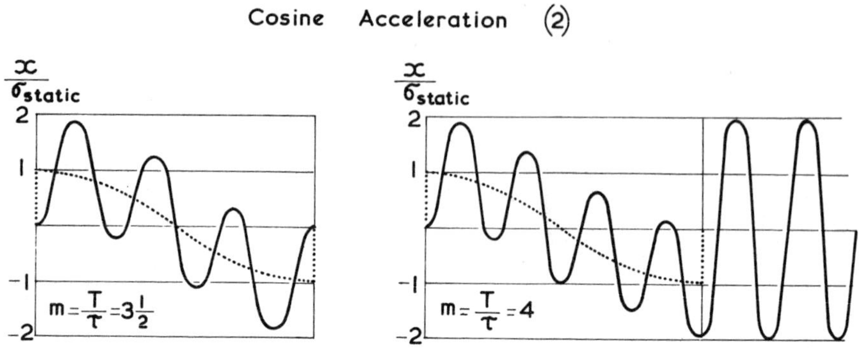
Cosine Acceleration (2)
x / σ static
2
1
-1
-2
m = T/τ = 3 1/2
m = T/τ = 4

Fig. I - 8

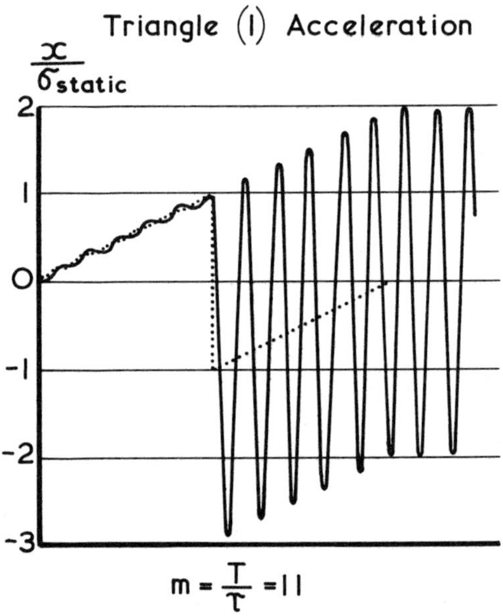
Triangle (1) Acceleration
x / σ static
2
1
0
-1
-2
-3
m = T/τ = 11

Fig. I - 9

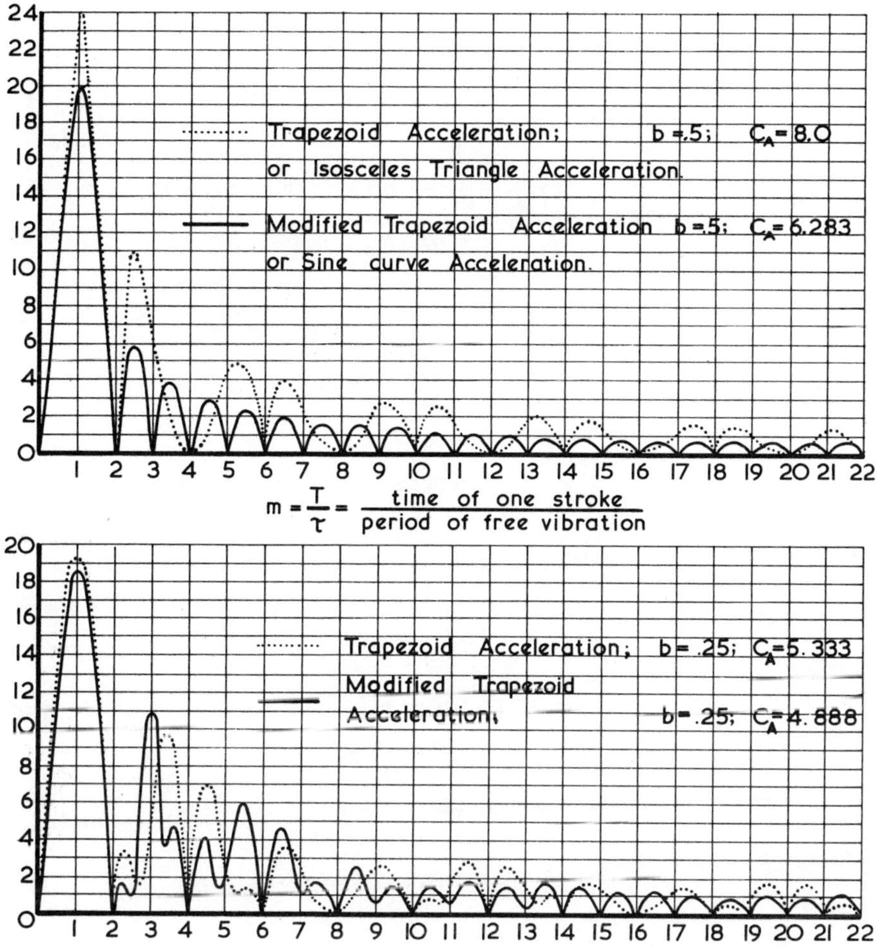

24
22
20
18
16
14
12
10
8
6
4
2
0
Trapezoid Acceleration; b =.5; C_A = 8.0
or Isosceles Triangle Acceleration.
Modified Trapezoid Acceleration b =.5; C_A = 6.283
or Sine curve Acceleration.
1 2 3 4 5 6 7 8 9 10 11 12 13 14 15 16 17 18 19 20 21 22
m = T/τ = time of one stroke / period of free vibration
20
18
16
14
12
10
8
6
4
2
0
Trapezoid Acceleration; b= .25; C_A=5.333
Modified Trapezoid Acceleration; b= .25; C_A=4.888
1 2 3 4 5 6 7 8 9 10 11 12 13 14 15 16 17 18 19 20 21 22

Fig. I - 10

Diagrams	Name	C_A	Conditions for min. and max. residue vibrational stresses	x/σ_{st} min,max	$C_A \frac{x}{\sigma_{st}}$
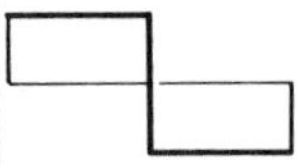	Uniform (3)	4	when m is an even number	0	
			when m is an odd number	± 4	± 16
	Cosine (2)	$\pm\frac{\pi^2}{2} = 4.935$	when m is a full number $+\frac{1}{2}$	0	
			when m is a full number	$\geq \pm 2$	$\geq \pm 9.87$
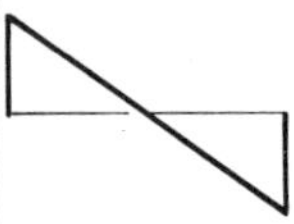	Triangle (2)	± 6.0	for all values of m	> 0	
			when m is close to a full number	$\geq \pm 2$	$\geq \pm 12$
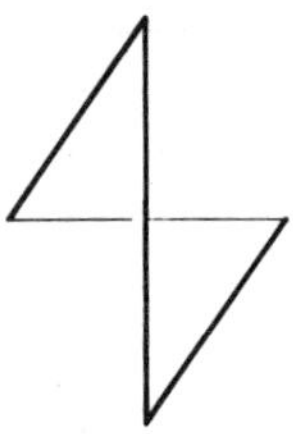	Triangle (1)	± 12	residue vibrations exist always		
			min. value	$\geq \pm 1.4$	$\geq \pm 16.8$
			max. value	$\geq \pm 2$	$\geq \pm 24$
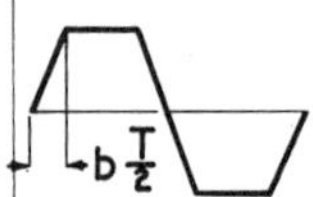	Trapezoid b=.25	± 5.33	when m = ~12, max. value		$\leq \pm 2.5$
	Trapezoid b=.5	± 8.0	when m = ~12, max. value		$\leq \pm 2$
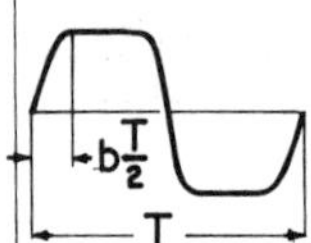	Modified Trapezoid b=.25	± 4.89	when m = ~12		$\leq \pm 1.7$
			≥ 20		$\leq \pm 1.0$
	Modified Trapezoid b=.5	± 6.28	when m = ~12		$\leq \pm 1.0$
			≥ 20		$\leq \pm .6$

Fig. I - 11

CHAPTER II

MECHANISMS

Properties of mechanisms were analyzed in a limited way; the analysis was applied only to mechanisms either used in automatic machines or considered for application. Though it was not a systematic research in the field of mechanisms, nevertheless three different groups of mechanisms were studied.

I. Mechanisms producing reciprocal motion:
   (a) Crank and slide;
   (b) Oscillating arm;
   (c) Double crank drive.

II. Mechanisms with varying angular velocity of the output shaft:
   (a) Elliptical gears;
   (b) Offset cranks (or crank and slot).

III. Mechanisms for intermittent motion:
   (a) Geneva drives—external or regular;
   (b) Geneva drives—internal;
   (c) Cam driven wheels.

Mechanisms III (c) are only described, but their properties are fully explained in the section of the book dealing with cams.

The results of the analysis are presented in equations and on the charts for a quick reference. The data give a complete picture of properties of mechanisms needed for the appraisal of them and for solutions of details of a machine, namely:

the displacement of the output;
the velocity, and
the acceleration.

The studies are made with an assumption that the driving (input) member rotates with uniform velocity. An analytic method is used. The basic approach is to write an equation which represents the displacement of the driven part as a function of the varying angle of rotation of the driving part. That angle can be expressed in terms of time t in seconds, due to the assumption of uniformity of the rotation of the driver; so actually, the time t is the variable in the equation.

Then, the first derivative of this displacement gives the equation of velocity in terms either inches per second in case of lineal displacement or radians per second in case of angular displacement.

The second derivative of the displacement equation is the equation of the acceleration either in inches per second per second or in radians per second per second.

The analysis starts with a schematic representation of the mechanism and the selection of an initial position of the driver. Next, it is assumed that the driver turns on a certain angle from the initial position and a preliminary equation for the displacement of the driven part is written.

The next step is to write auxiliary expressions describing geometric relation between the movements of the driving and the driven parts. These expressions are introduced into the preliminary equation, so eventually the displacement of the driven part is expressed in terms of rotation of the driver.

The final equation usually includes parameters, which describe physical dimensions of the mechanism, or, in most cases, ratios of these dimensions. Then the first and second derivatives of the displacement equation are the equation for the velocity and for the acceleration of the driven part.

Mechanisms have "built-in" kinematic characteristics, so the designer has little chance to improve them, but it is necessary to know what can be expected from them and, especially, their limitations.

Crank and Slide

(I_{a1}) The simplest case is presented by a "Scotch Yoke", it is equivalent to a crank drive with infinitely long connecting rod.

The initial position of the Crank Pin is at the point a; at that moment, the slide is in the farthest position from the shaft axis 0. When the crank pin moves from the position a to the position b, the crank arm turns on the angle β and the slide moves on the distance

$$\overline{ac} = s$$

It is evident that the displacement $s = R(1 - \text{Cos } \beta)$ (1)

The total stroke S of the slide is produced when $\beta = 180°$ or π radians, then

$$S = 2R \text{ or } R = \frac{S}{2} \qquad \text{(a)}$$

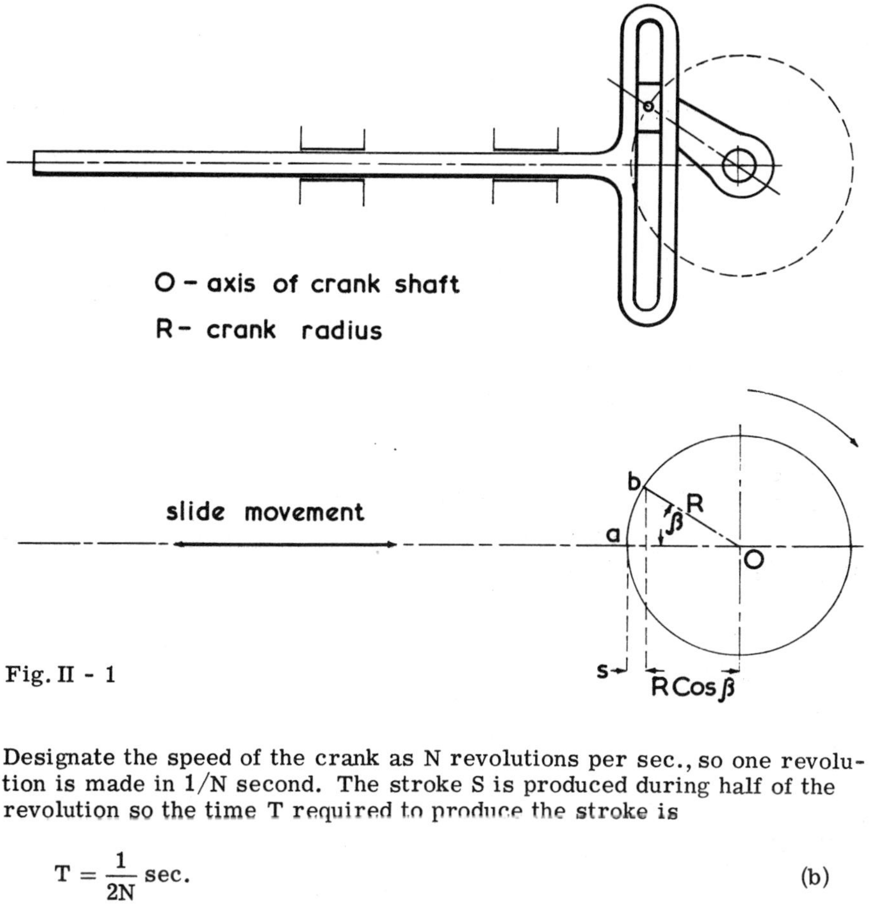

Fig. II - 1

Designate the speed of the crank as N revolutions per sec., so one revolution is made in 1/N second. The stroke S is produced during half of the revolution so the time T required to produce the stroke is

$$T = \frac{1}{2N} \text{ sec.} \qquad \text{(b)}$$

During the stroke S the angle β varies uniformly between 0 and π radians and the time varies also uniformly between 0 and T.

If the variable time is designated as t, the variation of the angle β can be presented as

$$\beta = \pi \frac{t}{T} \text{ radians} \qquad \text{(c)}$$

The same result can be obtained using the angular velocity of the crank

$$w = 2\pi N, \text{ so } \beta = wt$$

(1/T from the expression (b) is replaced by 2N in the equation (c)).

Introduce the auxiliary expressions (a) and (c) into the equation (1). Then the final equation for the displacement of the slide will be

$$s = S \frac{1 - \text{Cos}\ \pi \frac{t}{T}}{2} \qquad (2)$$

or in dimensionless form

$$\frac{s}{S} = \frac{1 - \text{Cos}\ \pi \frac{t}{T}}{2} = f_s(t) \qquad (2_1)$$

The equation (2_1) shows clearly that the variable function

$$f_s(t) \text{ or } f_s\left(\frac{t}{T}\right)$$

varies between 0 and unity

The velocity of the slide is represented by the first derivative of the equation (2)

$$v = \frac{ds}{dt} = \frac{S}{T} \cdot \frac{\pi}{2} \cdot \text{Sin}\ \pi \frac{t}{T} \qquad (3)$$

The ratio $\frac{S}{T}$ has dimensions $\frac{\text{inch}}{\text{sec.}}$, while other items are dimensionless, because dimension of the velocity is the same as the dimension of $\frac{S}{T}$.

$$\frac{v}{S/T} = \frac{\pi}{2} \cdot \text{Sin}\ \pi \frac{t}{T} = \frac{\pi}{2} \cdot f_v(t) \qquad (3_1)$$

The variable part of this equation $f_v(t) = \text{Sin}\ \pi \frac{t}{T}$ changes between 0 and 1, therefore, $\frac{\pi}{2}$ represents the maximum value of the ratio $\frac{v}{S/T}$, so it is the coefficient of maximum velocity

$$C_v = \frac{\pi}{2} = 1.571 \qquad (d)$$

The ratio $\frac{S}{T}$ represents the average velocity of the slide, so the maximum velocity during the stroke is C_v times greater.

The equation for the acceleration of the slide is either the first derivative of the velocity equation (3) or the second derivative of the displacement equation (2)

$$a = \frac{dv}{dt} = \frac{d^2s}{dt^2} = \frac{S}{T^2} \cdot \frac{\pi^2}{2} \cdot \text{Cos}\ \pi \frac{t}{T} = \frac{S}{T^2} \cdot \frac{\pi^2}{2} \text{Cos}\ \beta \qquad (4)$$

The ratio $\frac{S}{T^2}$ has dimensions of the acceleration, namely, inch/sec.^{2}, so the equation for the acceleration can be written in the dimensionless form

$$\frac{a}{S/T^2} = \frac{\pi^2}{2} \cdot \text{Cos}\ \pi\frac{t}{T} = \frac{\pi^2}{2} f_a(t) \qquad (4_1)$$

The variable part of this equation $f_a(t) = \text{Cos}\ \pi\frac{t}{T}$ varies between +1 and —1, therefore, $\frac{\pi^2}{2}$ represents the maximum acceleration coefficient

$$C_A = \pm\frac{\pi^2}{2} = \pm 4.935 \qquad (e)$$

Introduce into the equation (4) the expressions (a) and (b) and accept $f_a(t) = \pm 1$,

$$a_{max} = \pm\frac{\pi^2}{2} \cdot 2R \cdot (2N)^2 = \pm 4\pi^2 RN^2 \cong \pm 40 \cdot R \cdot N^2 = \pm 20 \cdot 2R \cdot N^2 \qquad (5)$$

This expression permits a quick appraisal of the maximum values of the acceleration.

In equation (4) the variable part has maximum values when t = 0 or T; these moments coincide with both ends of the stroke of the slide, so the diagram of the acceleration has the form of Cos β.

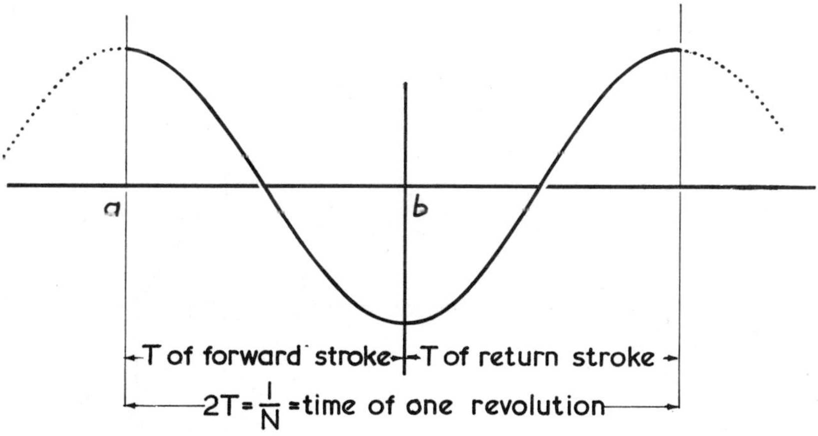

Fig. II - 2

This diagram is ideal because the changes of dynamic forces are gradual, but only in case of a reciprocal motion. When this drive is used for an intermittent motion, only the part from a to b is utilized, so at both ends of the stroke, the acceleration changes instantaneously between zero and maximum value, what is equivalent to a shock.

In case of lineal acceleration, it is convenient to express a_{max} in terms of gravity acceleration.

$$g = 386 \text{ inch/sec.}^2$$

Then equation (5) will be

$$\frac{a_{max}}{g} = \frac{39.4789}{386} RN^2 = 0.1023\ RN^2 \cong \frac{RN^2}{10} \cong \frac{SN^2}{20} \qquad (5_1)$$

Then to find the maximum dynamic force, this expression should be multiplied by the weight of parts moving on the distance S.

Scotch Yoke is not used very often: it is more expensive than connecting rods and there are difficulties to decrease clearances between the yoke and the sliding member, which caused a back-lash in the middle of each stroke.

It was analyzed to demonstrate step by step the procedure, so it would be possible to omit some detailed remarks in the analysis of more complicated mechanisms.

The only parameter which enters into equations is the radius of the crank.

(Ia_2) Crank — Connecting Rod — Slide

It can be presented schematically

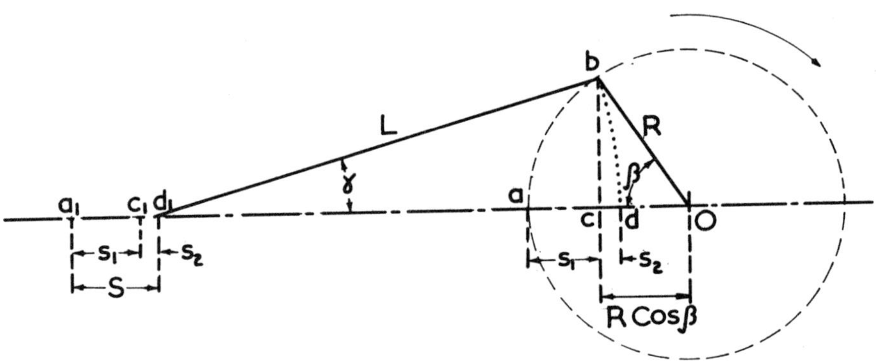

Fig. II - 3

The initial position of the crank pin is at the point a, at the moment the slide is at the point a_1. When the crank rotates on the angle β and the crank pin comes to the point b, the slide moves to the point d_1.

The displacement of the slide $a_1 d_1$ is caused by two factors:

(1) the rotation of the crank on the angle β and

(2) the movement of the connecting rod L on an angle γ.

$s_1 = R(1 - \text{Cos}\ \beta)$, as in case of Scotch Yoke and

$s_2 = L(1 - \text{Cos}\ \gamma)$, due to the presence of the connecting rod;

$$s = s_1 + s_2 = R(1 - \text{Cos}\ \beta) + L(1 - \text{Cos}\ \gamma) \qquad (1)$$

The triangles $d_1 bc$ and $0\,bc$ have a common side $\overline{bc}$.

$$\overline{bc} = R\ \text{Sin}\ \beta = L\ \text{Sin}\ \gamma$$

$$\text{Sin}\ \gamma = \frac{R}{L}\ \text{Sin}\ \beta$$

$$\text{Cos}\ \gamma = \sqrt{1 - \text{Sin}^2\ \gamma} = \sqrt{1 - \left(\frac{R}{L}\right)^2 \text{Sin}^2\ \beta} = \frac{R}{L}\sqrt{\left(\frac{L}{R}\right)^2 - \text{Sin}^2\ \beta} \qquad (a)$$

+ sign is used because angle γ is smaller than 90°.

During the analysis of Scotch Yoke, it was found

$$\beta = \pi \cdot \frac{t}{T} \qquad (b)$$

and

$$R = \frac{S}{2} \qquad (c)$$

Change the equation (1) by the introduction into it expressions (a), (b) and (c).

$$s = R\left[(1 - \text{Cos}\ \beta) + \frac{L}{R}(1 - \text{Cos}\ \gamma)\right] = S\left[\frac{1 - \text{Cos}\ \beta}{2} + \frac{L}{R} \cdot \frac{1 - \text{Cos}\ \gamma}{2}\right] =$$

$$= S\left[\frac{1 + \frac{L}{R}}{2} - \frac{\text{Cos}\ \beta}{2} - \frac{\frac{L}{R} \cdot \frac{R}{L} \cdot \sqrt{\left(\frac{L}{R}\right)^2 - \text{Sin}^2\ \beta}}{2}\right] =$$

$$= S\left[\frac{1 + \frac{L}{R}}{2} - \frac{\text{Cos}\ \pi \frac{t}{T}}{2} - \frac{\sqrt{\left(\frac{L}{R}\right)^2 - \text{Sin}^2\ \pi \frac{t}{T}}}{2}\right] \qquad (2)$$

or in dimensionless form, the displacement equation is

$$\frac{s}{S} = \frac{(1 - \text{Cos}\ \beta) + \frac{L}{R} - \sqrt{\left(\frac{L}{R}\right)^2 - \text{Sin}^2\ \beta}}{2} \tag{2_1}$$

It is shown on Chart 1.

The first derivative of the equation (2) gives the equation of the velocity of the slide

$$v = \frac{S}{T} \cdot \frac{\pi}{2}\ \text{Sin}\ \beta \left(1 + \frac{\text{Cos}\ \beta}{\sqrt{\left(\frac{L}{R}\right)^2 - \text{Sin}^2\ \beta}}\right) \tag{3}$$

In dimensionless form

$$\frac{v}{S/T} = \frac{\pi}{2}\ \text{Sin}\ \beta \left(1 + \frac{\text{Cos}\ \beta}{\sqrt{\left(\frac{L}{R}\right)^2 - \text{Sin}^2\ \beta}}\right) \tag{3_1}$$

$$\frac{S}{T} = 2R \div \frac{1}{2N} = 4RN; \qquad \frac{\pi}{2} \cdot \frac{S}{T} = 2\pi NR = wR$$

The chart 2 is drawn in a dimensionless form for different ratios $\frac{L}{R}$.

$$\frac{v}{2RN} = \pi\ \text{Sin}\ \beta \left(1 + \frac{\text{Cos}\ \beta}{\sqrt{\left(\frac{L}{R}\right)^2 - \text{Sin}^2\ \beta}}\right) \tag{3_2}$$

To receive the value of the velocity, the ordinates of the chart should be multiplied by 2RN. The derivative of the velocity equation is the equation of the acceleration

$$a = \frac{dv}{dt} = \frac{S}{T^2} \cdot \frac{\pi^2}{2} \left[\text{Cos}\ \beta + \frac{\left(\frac{L}{R}\right)^2 . \text{Cos}^2\ \beta}{\left[\left(\frac{L}{R}\right)^2 - \text{Sin}^2\ \beta\right]^{3/2}} - \frac{\text{Sin}^2\ \beta}{\sqrt{\left(\frac{L}{R}\right)^2 - \text{Sin}^2\ \beta}} \right] \tag{4}$$

$$\frac{S}{T^2} \cdot \frac{\pi^2}{2} = 2R\ .\ 4N^2\ .\ \frac{\pi^2}{2} = 2\pi^2\ .\ 2RN^2$$

In the dimensionless form, the equation of the acceleration can be written as

$$\frac{a}{2RN^2} = 2\pi^2 \left[\text{Cos}\ \beta + \frac{\left(\frac{L}{R}\right)^2 \text{Cos}^2\ \beta}{\left(\sqrt{\left(\frac{L}{R}\right)^2 - \text{Sin}^2\ \beta}\right)^3} - \frac{\text{Sin}^2\ \beta}{\sqrt{\left(\frac{L}{R}\right)^2 - \text{Sin}^2\ \beta}} \right] \qquad (4_1)$$

In this form, the acceleration is presented on the chart 3 for different values of the ratio $\frac{L}{R}$ and to receive the values of the acceleration, the ordinates of the chart should be multiplied by $2RN^2$.

In the equation (4) the first expression $\frac{S}{T^2} \cdot \frac{\pi^2}{2} \cdot \text{Cos}\ \beta$ represents the acceleration of Scotch Yoke; two other items show the influence of the ratio $\frac{L}{R}$.

When $\frac{L}{R} = 5$ or larger, the maximum acceleration is at both ends of the stroke, or when $\beta = 0°$ or $180°$.

Designate the maximum acceleration of Scotch Yoke as A_{SY}

When $\beta = 0°$, $\quad \frac{a_{max}}{A_{SY}} = 1 + \frac{R}{L}$;

When $\beta = 180°$, $\quad \frac{a_{max}}{A_{SY}} = -\left(1 - \frac{R}{L}\right)$.

It means—the magnitude of the acceleration at two ends of the stroke, when $\frac{L}{R} \geqslant 5$, is

$$a_{max} = A_{SY}\left(1 \pm \frac{R}{L}\right).$$

It appears that the distortion is minor, but the difference in acceleration can be quite noticeable. For instance, for $\frac{R}{L} = 0.2$ the accelerations at two ends of the stroke are proportional to 0. 8 and to 1. 2; in this case, the difference is 50%. In some cases, the distortion may be favorable: when a punch press has the crank shaft above the die, the inertia of the ram is the highest at the bottom of the work stroke, so it decreases the work load on the crank shaft and crank pin bearings.

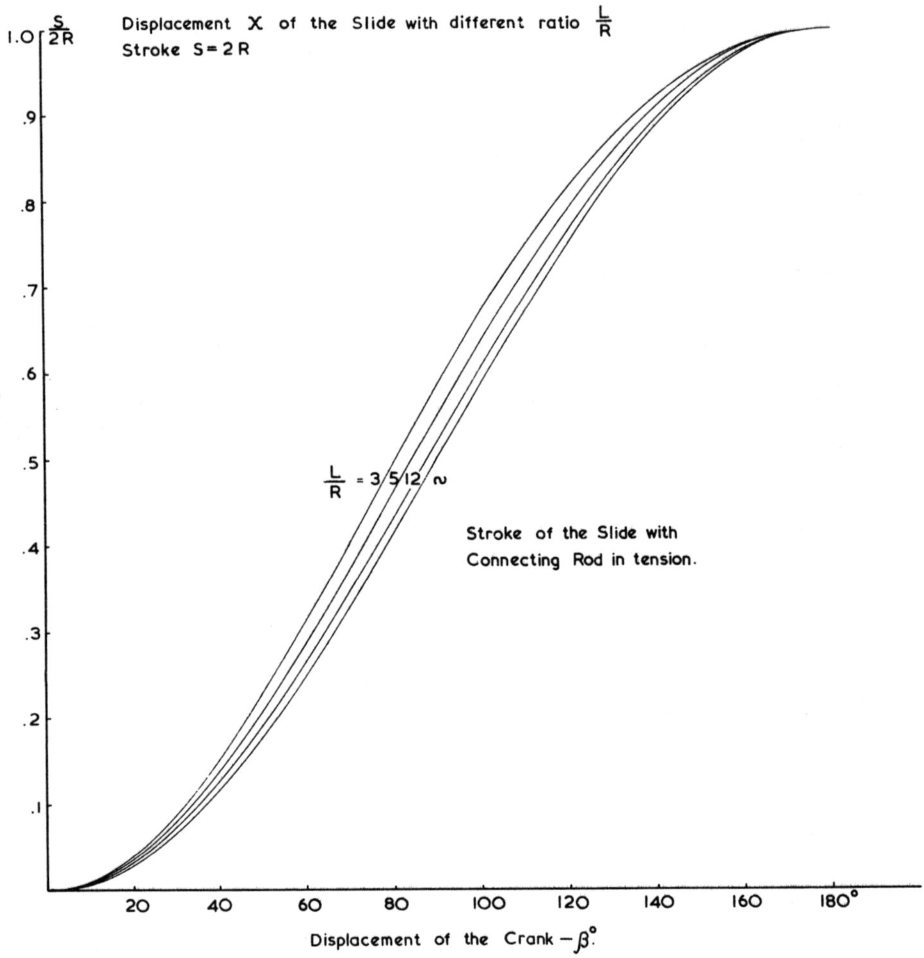

Chart 1. Crank and Slide Mechanism

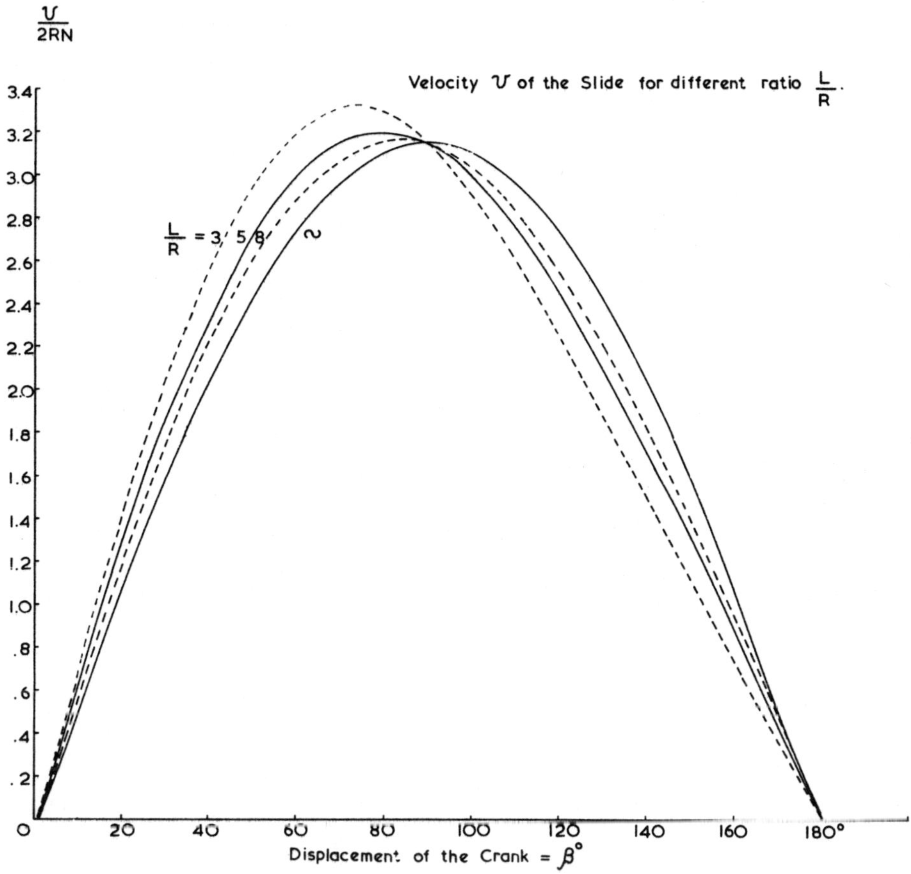

Chart 2. Crank and Slide Mechanism

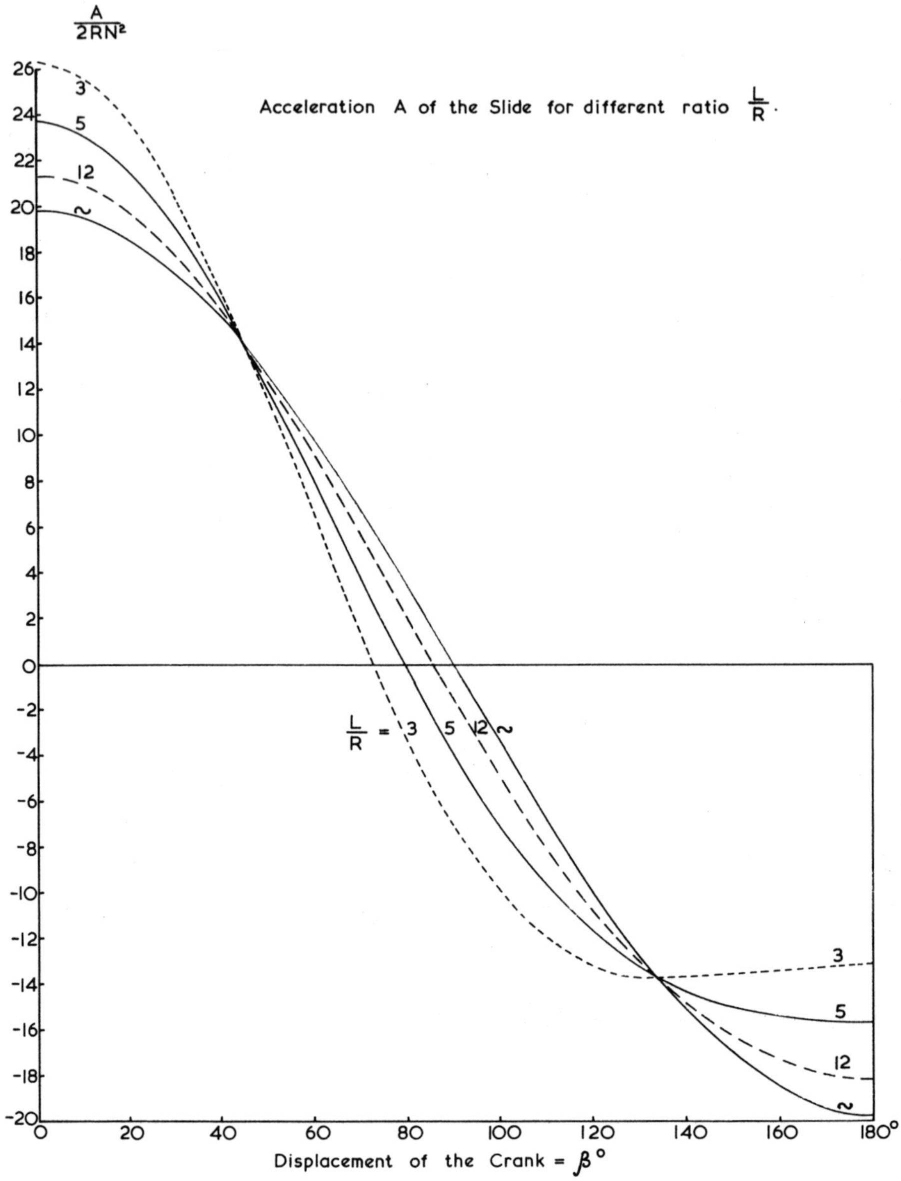

Chart 3. Crank and Slide Mechanism

(I_b) <u>Oscillating Arm—driven by crank</u>

O — center of rotation of the crank;
C — pivot point of the oscillating arm;
S = $\overline{OC}$ — Center to center distance;
r — radius of the crank

When the crank pin travels over the arc ADE, the driver arm moves slower. When the crank pin travels over the arc EFA, the arm moves faster.

The crank pin slides in a slot in the oscillating arm.

The sketch on Fig. II-4 is used for the analysis of the slow stroke. The initial position of the crank pin is at the point A and the pin moves clockwise.

When the pin is at the point B, the crank rotated on the angle β from the initial position at the point A. During that time the arm moves on the angle γ.

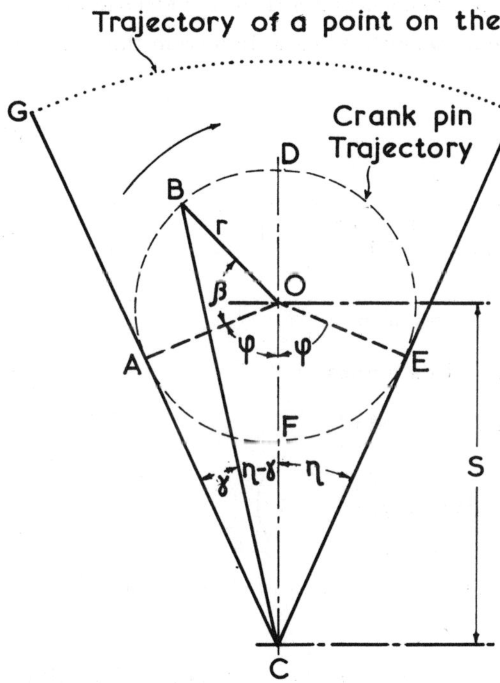

Fig. II - 4

During the slow stroke, the angle β varies from 0° to 360°-2φ°, where 2φ° is the rotation of the crank during the fast stroke. During that time, the angle γ° varies from 0° to 2η°, where 2η is the maximum angular movement of the arm.

Oscillating movement of the arm can be only in cases when $S > r$; when $S < r$, the arm rotates in the same direction as the crank but with varying angular velocity (this case is analyzed as case #II b of "Offset Cranks".) The problem is to express the angle γ as a function of the angle β, which varies uniformly.

The triangle COB has two known sides: $\overline{OC} = S$ and $\overline{OB} = r$; also two known angles: $\angle COB = \beta + \varphi$ and $\angle OCB = \eta - \gamma$.
Then the third angle, located opposite the side $\overline{OC}$, is $180° - (\beta + \varphi) - (\eta - \gamma) = 180° - (\beta + \varphi + \eta - \gamma)$ (a)

The property of the triangle permits to write

$$\frac{r}{\text{Sin}(\eta - \gamma)} = \frac{S}{\text{Sin}[180 - (\beta + \varphi) - (\eta - \gamma)} = \frac{S}{\text{Sin}[(\beta + \varphi) + (\eta - \gamma)]} \quad \text{(b)}$$

It is necessary to eliminate the angles φ and η from the equation (a).

The geometry of the mechanism permits to establish from the triangle COA

$$\text{Cos}\,\varphi = \text{Sin}\,\eta = \frac{r}{S} \quad \text{(c)}$$

$$\text{Sin}\,\varphi = \text{Cos}\,\eta = \sqrt{1 - \left(\frac{r}{S}\right)^2} = \frac{r}{S} \cdot \sqrt{\left(\frac{S}{r}\right)^2 - 1} \quad \text{(d)}$$

Transform Sine values in the equation (b)

$$\text{Sin}(\eta - \gamma) = \text{Sin}\,\eta\,\text{Cos}\,\gamma - \text{Cos}\,\eta\,\text{Sin}\,\gamma$$

introduce values from equations (c) and (d)

$$\text{Sin}(\eta - \gamma) = \frac{r}{S}\left(\text{Cos}\,\gamma - \text{Sin}\,\gamma \cdot \sqrt{\left(\frac{S}{r}\right)^2 - 1}\right) \quad \text{(e)}$$

$$\text{Sin}[(\beta + \varphi) + (\eta - \gamma)] = \text{Sin}(\beta + \varphi) \cdot \text{Cos}(\eta - \gamma) + \text{Cos}(\beta + \varphi) \cdot \text{Sin}(\eta - \gamma) =$$

$$= (\text{Sin}\,\beta \cdot \text{Cos}\,\varphi + \text{Cos}\,\beta \cdot \text{Sin}\,\varphi)(\text{Cos}\,\eta \cdot \text{Cos}\,\gamma + \text{Sin}\,\eta\,\text{Sin}\,\gamma) + (\text{Cos}\,\beta \cdot \text{Cos}\,\varphi - \text{Sin}\,\beta \cdot \text{Sin}\,\varphi)(\text{Sin}\,\eta\,\text{Cos}\,\gamma - \text{Cos}\,\eta\,\text{Sin}\,\gamma)$$

Again, introduce values from equations (c) and (d).

Then, after the transformations

$$\text{Sin}\,[(\beta + \varphi) + (\eta - \gamma)] = \text{Cos}\,\beta\,\text{Cos}\,\gamma + \text{Sin}\,\beta\,\text{Sin}\,\gamma \qquad \text{(f)}$$

Introduce the expressions (e) and (f) into the equation (b)

$$\frac{S}{r}\,\frac{r}{\text{Cos}\,\gamma - \text{Sin}\,\gamma\sqrt{\left(\frac{S}{r}\right)^2 - 1}} = \frac{S}{\text{Cos}\,\beta\,\text{Cos}\,\gamma + \text{Sin}\,\beta\,\text{Sin}\,\gamma};$$

$$\text{Cos}\,\gamma - \text{Sin}\,\gamma\sqrt{\left(\frac{S}{r}\right)^2 - 1} = \text{Cos}\,\beta\,\text{Cos}\,\gamma + \text{Sin}\,\beta\,\text{Sin}\,\gamma;$$

$$\text{Sin}\,\gamma\left(\text{Sin}\,\beta + \sqrt{\left(\frac{S}{r}\right)^2 - 1}\right) + \text{Cos}\,\gamma\,(\text{Cos}\,\beta - 1) = 0$$

Divide by Cos γ,

$$\text{tg}\,\gamma\left(\text{Sin}\,\beta + \sqrt{\left(\frac{S}{r}\right)^2 - 1}\right) = 1 - \text{Cos}\,\beta$$

Designate the angle γ as γ_s to show that it belongs to the slow stroke

$$\tan \gamma_s = \frac{1 - \text{Cos}\,\beta}{\text{Sin}\,\beta + \sqrt{\left(\frac{S}{r}\right)^2 - 1}} \qquad (1)$$

The angular displacement of the oscillating arm should be presented as

$$\gamma_s = \text{arc tan}\,\frac{1 - \text{Cos}\,\beta}{\text{Sin}\,\beta + \sqrt{\left(\frac{S}{r}\right)^2 - 1}} \qquad (2)$$

The lineal displacement of a point on the arm $\overline{CH} = R$

$$f_s = R \,.\, \gamma_s = R \,.\, f(\beta)$$

The lineal stroke is expressed as F when $\beta = 2\pi - 2\varphi$ or $\beta° = 360° - 2\varphi°$.

The chart 4 shows a dimensionless ratio $\frac{f_s}{F}$ and also $\frac{f_f}{F}$ for the complete one revolution of the crank.

The displacement of the arm is affected by the ratio $\frac{S}{r}$, which can be selected by the designer. The selection affects the relation between the time of the slow stroke and the time of the fast stroke; also, it affects the value of the angle 2η (see equation (c)).

The crank rotates at uniform speed, therefore, the time of the slow stroke is proportional to $360° - 2\varphi°$, and the time of the fast stroke is proportional to $2\varphi°$.

So,

$$\frac{\text{Time of slow stroke}}{\text{Time of fast stroke}} = \frac{360 - 2\varphi°}{2\varphi} = \frac{180 - \varphi°}{\varphi°}$$

Dynamic forces (acceleration) are inversely proportional to the square of time, therefore

$$\frac{\text{Dynamic forces during fast stroke}}{\text{Dynamic forces during slow stroke}} = \frac{180° - \varphi°}{\varphi°}^{2}$$

Example 1. $\frac{S}{r} = 1.5;$

from equation (c) $\quad \text{Cos } \varphi = \frac{r}{S} = .(6)\ldots$ and $\varphi° = 48°.19$

$$\left(\frac{180 - 48.19}{48.19}\right)^2 = 2.73^2 = 7.45$$

Dynamic forces during the fast stroke will be almost 7 . 5 times greater than during the slow stroke.

Example 2. $\frac{S}{r} = 2; \text{Cos } \varphi = 0.5; \varphi° = 60°$

$$\left(\frac{180 - 60}{60}\right)^2 = 2^2 = 4$$

Dynamic forces during the return stroke are four times greater.

Example 3. $\frac{S}{r} = 2.5; \text{Cos } \varphi = 0.4; \varphi° = 66.422$

$$\left(\frac{180 - 66.422}{66.422}\right)^2 = 1.71^2 = 2.924$$

So dynamic forces during the fast stroke are less than three times the dynamic forces during the slow stroke.

The next step in the analysis of the mechanism is to take first and second derivative of the displacement equation (2). In this case, the limits of the angle β vary when the ratio $\frac{S}{r}$ is changed, so the connection between the angle β and the speed of the crank is different than in the first two mechanisms.

If the speed of the crank is N RPS; the time of one revolution $= \frac{1}{N}$ sec.

The time t required to rotate the crank on the angle β is

$$t = \frac{\beta}{2\pi} \cdot \frac{1}{N}; \text{therefore } \beta = 2\pi Nt$$

It should be remembered that during the differentiation, the angle β should be expressed in radians, but in the final equation, for convenience of making calculations, it can be written as $\beta°$.

The angular velocity for the slow stroke

$$W_s = \frac{d(\arc \tan \gamma)}{dt} = \frac{dx}{dt(1 + x^2)}, \text{ where } x = \frac{1 - \cos 2\pi Nt}{\sin 2\pi Nt + \sqrt{\left(\frac{S}{r}\right)^2 - 1}}.$$

$$\frac{dx}{dt} = 2\pi N \frac{\sin^2 2\pi Nt + \sin 2\pi Nt\sqrt{\left(\frac{S}{r}\right)^2 - 1} - \cos 2\pi Nt + \cos^2 2\pi Nt}{\left(\sin 2\pi Nt + \sqrt{\left(\frac{S}{r}\right)^2 - 1}\right)^2}$$

For convenience, replace $2\pi Nt$ by β under Sin and Cos.

$$\frac{dx}{dt} = 2\pi N \frac{1 - \cos \beta + \sin \beta\sqrt{\left(\frac{S}{r}\right)^2 - 1}}{\left(\sin \beta + \sqrt{\left(\frac{S}{r}\right)^2 - 1}\right)^2}; \qquad \text{(h)}$$

$$1 + x^2 = \frac{1 + \left(\frac{S}{r}\right)^2 - 2 \cos \beta + 2 \sin \beta\sqrt{\left(\frac{S}{r}\right)^2 - 1}}{\left(\sin \beta + \sqrt{\left(\frac{S}{r}\right)^2 - 1}\right)^2} \qquad \text{(i)}$$

Introduce (h) and (i) into the expression of the angular velocity; the denominators in the expression for $\frac{dx}{dt}$ and for $(1 + x^2)$ will be canceled.

$$W_s = 2\pi N \frac{1 - \cos \beta + \sin \beta\sqrt{\left(\frac{S}{r}\right)^2 - 1}}{1 - 2 \cos \beta + 2 \sin \beta\sqrt{\left(\frac{S}{r}\right)^2 - 1} + \left(\frac{S}{r}\right)^2}$$

For easier calculations, introduce +1 −1 into the denominator

$$W_s = 2\pi N \frac{1 - \cos\beta + \sin\beta\sqrt{\left(\frac{S}{r}\right)^2 - 1}}{2\left(1 - \cos\beta + \sin\beta\sqrt{\left(\frac{S}{r}\right)^2 - 1}\right) + \left[\left(\frac{S}{r}\right)^2 - 1\right]} \tag{3}$$

This equation is presented on the charts in the form

$$\frac{W_s}{N} = 2\pi \frac{1 - \cos\beta + \sin\beta\sqrt{\left(\frac{S}{r}\right)^2 - 1}}{2\left(1 - \cos\beta + \sin\beta\sqrt{\left(\frac{S}{r}\right)^2 - 1}\right) + \left[\left(\frac{S}{r}\right)^2 - 1\right]} \tag{3_1}$$

It is evident that the highest angular velocity is at the moment when the arm passes the point D, or when $\beta = 180° - \varphi°$. Then

$$\sin\beta = \sin(180 - \varphi°) = \sin\varphi = \frac{r}{S}\sqrt{\left(\frac{S}{r}\right)^2 - 1}$$

$$\text{and } \cos\beta = -\cos\varphi = -\frac{r}{S}.$$

and

$$\frac{W_s}{N} = 2\pi \frac{1 + \frac{r}{s} + \frac{r}{s}\left[\left(\frac{S}{r}\right)^2 - 1\right]}{2\left[1 + \frac{r}{s} + \frac{r}{s}\left(\frac{S^2}{r^2} - 1\right)\right] + \left(\frac{s}{r}\right)^2 - 1} = \frac{2\pi}{1 + \frac{s}{r}} \tag{3_2}$$

The angular acceleration

$$\Omega_s = \frac{dw_s}{dt} = 2\pi N \frac{d}{dt}\left(\frac{1 - \cos\beta + \sin\beta\sqrt{\left(\frac{S}{r}\right)^2 - 1}}{2\left(1 - \cos\beta + \sin\beta\sqrt{\left(\frac{S}{r}\right)^2 - 1}\right) + \left(\frac{S}{r}\right)^2 - 1}\right) =$$

$$= (2\pi N)^2 \frac{\left[\left(\frac{S}{r}\right)^2 - 1\right]\left(\sin\beta + \cos\beta\sqrt{\left(\frac{S}{r}\right)^2 - 1}\right)}{\left[2\left(1 - \cos\beta + \sin\beta\sqrt{\left(\frac{S}{r}\right)^2 - 1}\right) + \left(\frac{S}{r}\right)^2 - 1\right]^2} \tag{4}$$

The chart 6 is prepared for

$$\frac{\Omega_s}{N^2} = (2\pi)^2 \frac{\left[\left(\frac{S}{r}\right)^2 - 1\right]\left(\sin\beta + \cos\beta\sqrt{\left(\frac{S}{r}\right)^2 - 1}\right)}{\left[2\left(1 - \cos\beta + \sin\beta\sqrt{\left(\frac{S}{r}\right)^2 - 1}\right) + \left(\frac{S}{r}\right)^2 - 1\right]^2} \tag{4_1}$$

The equation can be checked: when the angular velocity has its maximum, the acceleration $\Omega_S = 0$.

It happens when $\beta = 180° - \varphi°$.

$$\text{Then } \sin\beta = \sin\varphi = \frac{r}{S}\sqrt{\left(\frac{S}{r}\right)^2 - 1};\ \cos\beta = -\cos\varphi = -\frac{r}{S}$$

When these values are introduced into the numerator of the equation (4).

$$\sin\beta + \cos\beta\sqrt{\left(\frac{S}{r}\right)^2 - 1} = \frac{r}{S}\sqrt{\left(\frac{S}{r}\right)^2 - 1} - \frac{r}{S}\sqrt{\left(\frac{S}{r}\right)^2 - 1} = 0$$

Fast Stroke

In the analysis of the fast stroke, the initial position of the crank is when the crank pin is at the point E. So, the angle β is measured from the radius $\overline{OE}$.

When charts 4, 5, and 6 are prepared for one revolution of the crank, the values of the angle β are added to the rotation of the crank during the slow stroke, which is equal to $(360 - 2\varphi)°$.

The procedure is the same as in the case of the slow stroke.

The triangle COB has two known sides:

$$\overline{OC} = S \text{ and } \overline{OB} = r;$$

also two known angles:

$$\angle\, COB = \varphi - \beta \text{ and } \angle\, OCB = \eta - \gamma$$

The angle opposite the side S is

$$180° - [(\varphi - \beta) + (\eta - \gamma)]° \tag{a}$$

The property of the triangle permits to write

$$\frac{r}{\sin(\eta - \gamma)} = \frac{S}{\sin[(\varphi - \beta) + (\eta - \gamma)]} \tag{b}$$

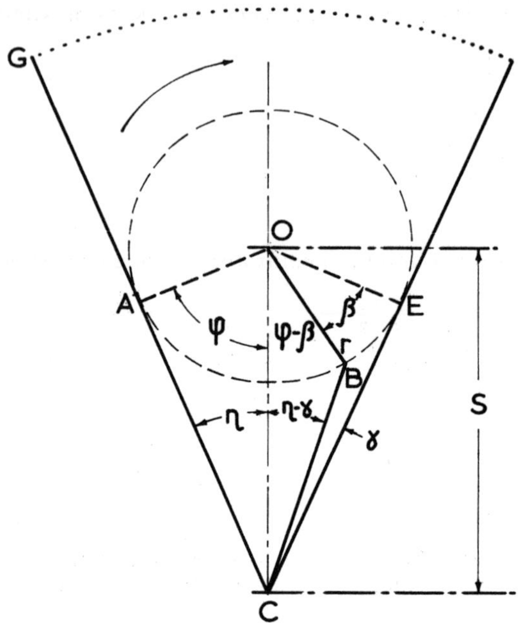

Fig. II - 5

The geometry of the mechanism gives the following relations

$$\text{Cos } \varphi = \text{Sin } \eta = \frac{r}{S} \tag{c}$$

$$\text{Sin } \varphi = \text{Cos } \eta = \frac{r}{S}\sqrt{\left(\frac{S}{r}\right)^2 - 1} \tag{d}$$

The transformation of Sin $(\eta - \gamma)$ and of Sin $[(\varphi - \beta) + (\eta - \gamma)]$ for the equation (b) and the elimination of angles φ and η are the same as in the case of the slow stroke. As the result

$$\tan \gamma_f = \frac{1 - \text{Cos } \beta}{\sqrt{\left(\frac{S}{r}\right)^2 - 1}} \tag{1}$$

The angular displacement of the arm

$$\gamma_f = \text{arc tan } \frac{1 - \text{Cos } \beta}{\sqrt{\left(\frac{S}{r}\right)^2 - 1}} \tag{2}$$

The angular velocity of the arm is found following the same procedure

$$w_f = 2\pi N \frac{\sin\beta\sqrt{\left(\frac{S}{r}\right)^2 - 1} - (1 - \cos\beta)}{\left[\left(\frac{S}{r}\right)^2 - 1\right] - 2\left[\sin\beta\sqrt{\left(\frac{S}{r}\right)^2 - 1} - (1 - \cos\beta)\right]} \quad (3)$$

The chart 5 is made for

$$\frac{w_f}{N} = 2\pi \frac{\sin\beta\sqrt{\left(\frac{S}{r}\right)^2 - 1} - (1 - \cos\beta)}{\left[\left(\frac{S}{r}\right)^2 - 1\right] - 2\left[\sin\beta\sqrt{\left(\frac{S}{r}\right)^2 - 1} - (1 - \cos\beta)\right]} \quad (3_1)$$

The maximum angular velocity is at the moment when $\beta = \varphi$, therefore

$$\sin\beta = \sin\varphi = \frac{r}{S}\sqrt{\left(\frac{S}{r}\right)^2 - 1} \text{ and } \cos\beta = \cos\varphi = \frac{r}{S}.$$

$$\text{and max. } w_f = \frac{2\pi N}{\frac{S}{r} - 1} \quad (3_2)$$

Ratio of maximum angular velocities during two strokes is

$$\frac{\max V_f}{\max V_c} = \frac{1 + \frac{S}{r}}{\frac{S}{r} - 1} = \frac{S + r}{S - r} \quad (3_3)$$

The angular acceleration during the fast stroke

$$\Omega_f = \frac{dw_f}{dt} = (2\pi N)^2 \frac{\left[\left(\frac{S}{r}\right)^2 - 1\right] . \left(\cos\beta . \sqrt{\left(\frac{S}{r}\right)^2 - 1} - \sin\beta\right)}{\left\{\left[\left(\frac{S}{r}\right)^2 - 1\right] - 2\left[\sin\beta\sqrt{\left(\frac{S}{r}\right)^2 - 1} - (1 - \cos\beta)\right]\right\}^2} \quad (4)$$

The chart 6 is drawn for

$$\frac{\Omega_f}{N^2}$$

Under some conditions, it may be desirable to have equations for lineal movements of a point on the oscillating arm on the distance L from the pivot point C.

$$CH = CG = L$$

The lineal displacement, measured along the arc GH.

$$f = \gamma \, . \, L \qquad (L_1)$$

where γ is in radians and determined from equation (1) for tan γ_s and tan γ_f. The full stroke F is produced when $\gamma = 2\eta$, so

$$F = 2\eta L \text{ and } \frac{F}{2\eta L} = 1$$

Then

$$f = \gamma L \frac{F}{2\eta L} = F \frac{\gamma}{2\eta}$$

In dimensionless form $\frac{f}{F} = \frac{\gamma}{2\eta}$

To convert the equations for velocity and acceleration, it is necessary to multiply the equations for angular velocity and acceleration by L and, to eliminate L, to multiply also by $\frac{F}{2\eta L}$; so actually the equations should be

multiplied by $\frac{F}{2\eta}$.

$$\left.\begin{aligned} v_s &= \frac{FN}{2\eta} \, . \, 2\pi \frac{\operatorname{Sin}\beta\sqrt{\left(\frac{S}{r}\right)^2 - 1} + (1 - \operatorname{Cos}\beta)}{\left[\left(\frac{S}{r}\right)^2 - 1\right] + 2\left[\operatorname{Sin}\beta\sqrt{\left(\frac{S}{r}\right)^2 - 1} + (1 - \operatorname{Cos}\beta)\right]} \\ v_f &= \frac{FN}{2\eta} \, . \, 2\pi \frac{\operatorname{Sin}\beta\sqrt{\left(\frac{S}{r}\right)^2 - 1} - (1 - \operatorname{Cos}\beta)}{\left[\left(\frac{S}{r}\right)^2 - 1\right] - 2\left[\operatorname{Sin}\beta\sqrt{\left(\frac{S}{r}\right)^2 - 1} - (1 - \operatorname{Cos}\beta)\right]} \end{aligned}\right\} \quad (L_3)$$

Lineal acceleration

$$\left.\begin{aligned} a_s &= \frac{FN^2}{2\eta}.\,(2\pi)^2 \frac{\left[\left(\frac{S}{r}\right)^2 - 1\right]\left(\text{Cos}\,\beta\sqrt{\left(\frac{S}{r}\right)^2 - 1} + \text{Sin}\,\beta\right)}{\left\{\left[\left(\frac{S}{r}\right)^2 - 1\right] + 2\left[\text{Sin}\,\beta\sqrt{\left(\frac{S}{r}\right)^2 - 1} + (1 - \text{Cos}\,\beta)\right]\right\}^2} \\ a_f &= \frac{FN^2}{2\eta}.\,(2\pi)^2 \frac{\left[\left(\frac{S}{r}\right)^2 - 1\right]\left(\text{Cos}\,\beta\sqrt{\left(\frac{S}{r}\right)^2 - 1} - \text{Sin}\,\beta\right)}{\left\{\left[\left(\frac{S}{r}\right)^2 - 1\right] - 2\left[\text{Sin}\,\beta\sqrt{\left(\frac{S}{r}\right)^2 - 1} - (1 - \text{Cos}\,\beta)\right]\right\}^2} \end{aligned}\right\} (L_4)$$

It should be remembered that the angle β is measured from different initial positions of the crank in case of the slow and of the fast strokes. When diagrams are drawn, the angle β, used in the equations of the fast stroke, is added to the maximum value of β at the end of the slow stroke.

Ic. Double Crank Drive

Two cranks drive a lever I-II which oscillates around the point P.

The arm with the point P moves in a straight line.

To simplify the analysis:

1. Neglect the influence of connecting rods between crank pins and points I and II—consider cranks as Scotch Yokes;
2. Consider that points I and II move along straight lines.

A special case is analyzed:

the crank driving the point I rotates at N RPS;
the crank driving the point II rotates at 2 N RPS.

Accept as initial positions when both crank pins, A_1 and A_2 are in their lowest positions. When the crank pin moves from A_1 to B_1, the crank arm turns on the angle β and the other crank rotates on the angle 2β. Radii of cranks are R and r.

The point I moves on the distance

$$f_1 = R(1 - \text{Cos}\,\beta) \tag{a}$$

and the point II moves on

$$f_2 = r(1 - \text{Cos}\,2\beta) \tag{b}$$

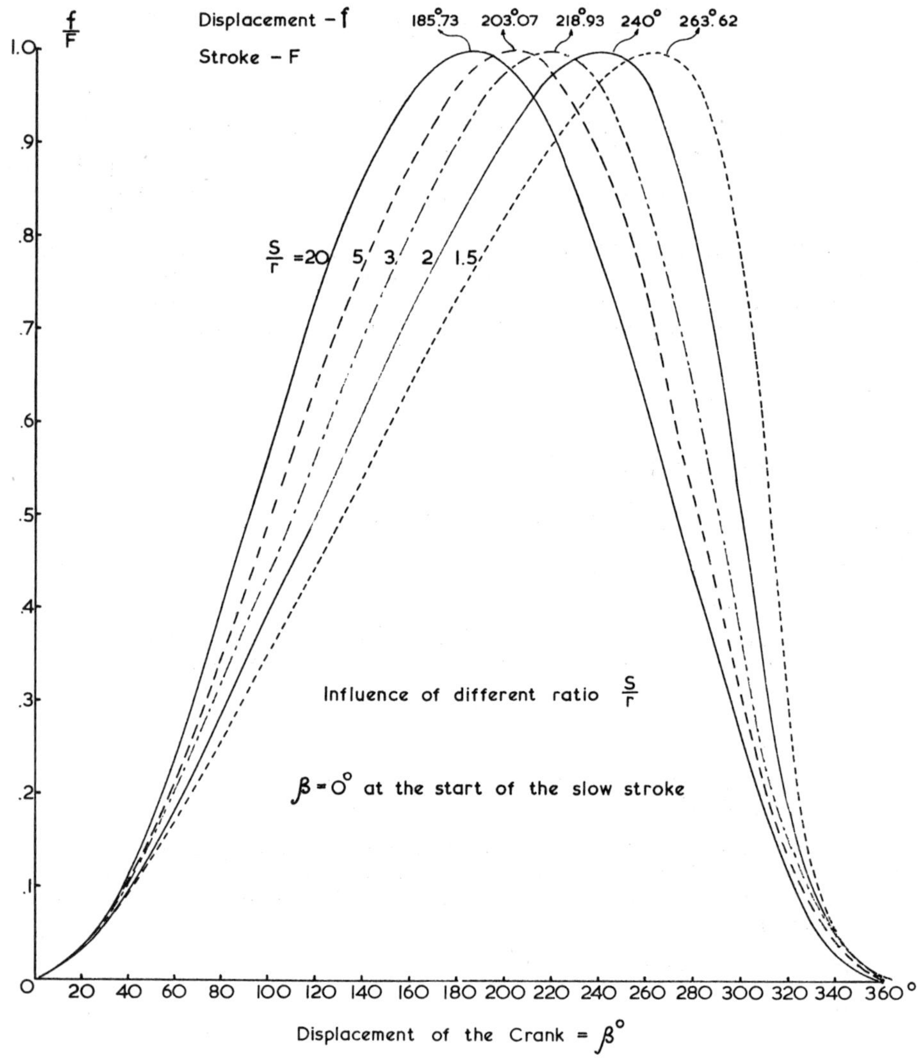

Chart 4. Oscillating Arm

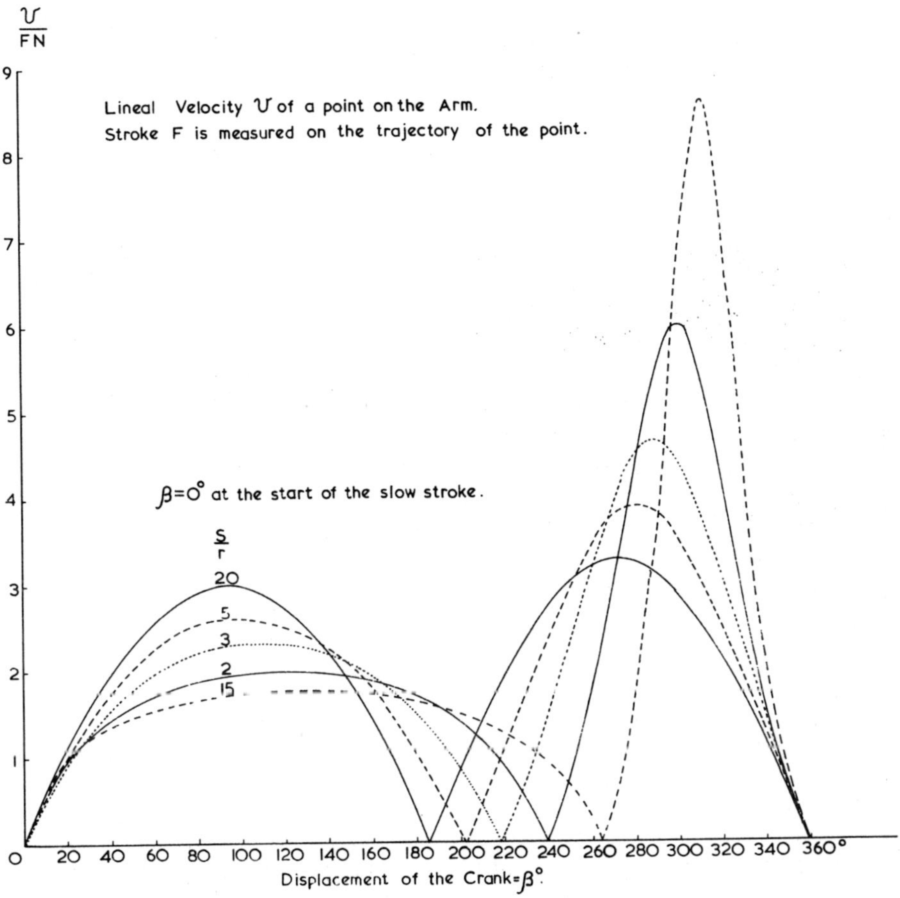

Chart 5. Oscillating Arm

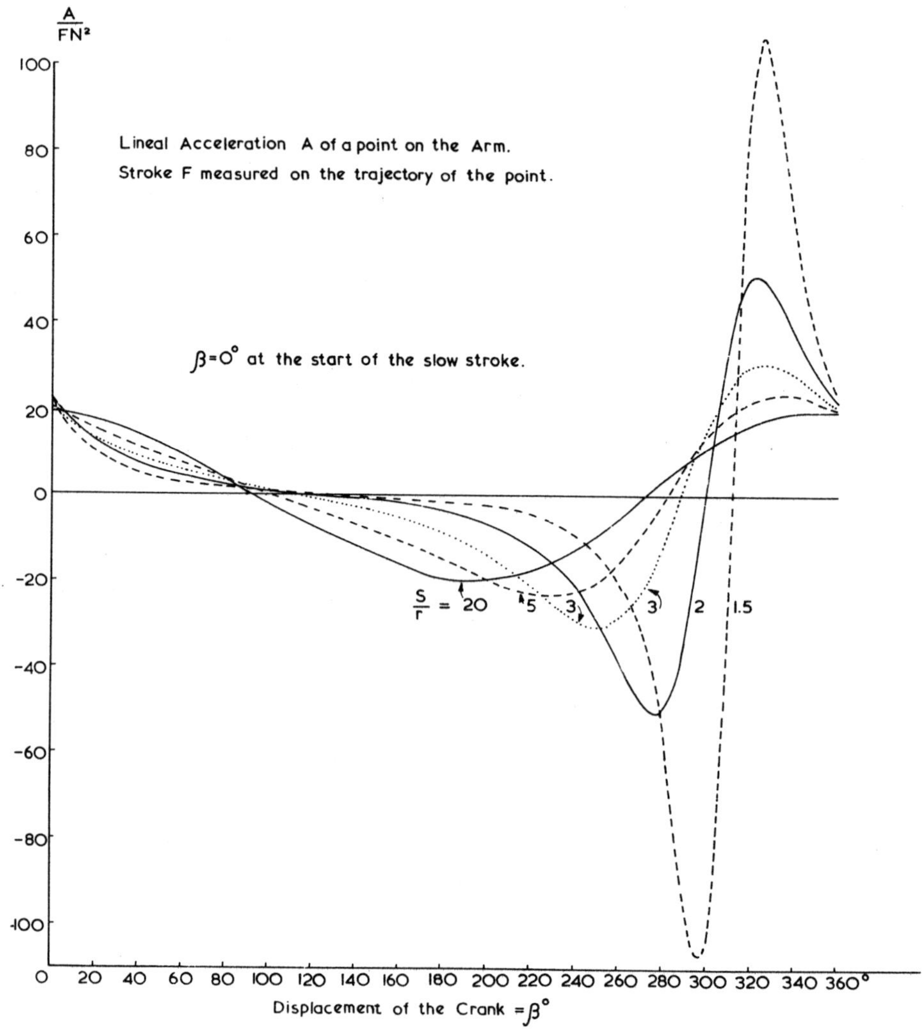

Chart 6. Oscillating Arm

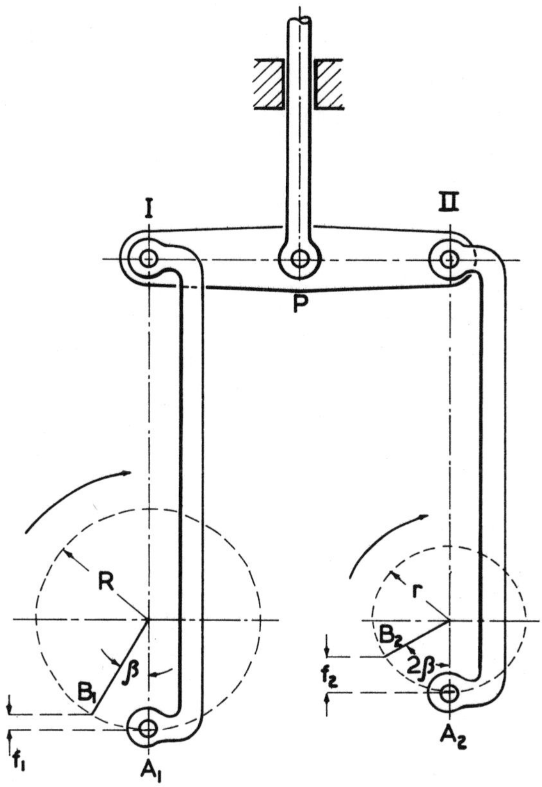

Fig. II - 6

Then the displacement of the point P will be the algebraic sum

$$f = \frac{f_1}{2} + \frac{f_2}{2} = \frac{R + r}{2} - \frac{R \operatorname{Cos} \beta + r \operatorname{Cos} 2\beta}{2} \tag{1}$$

Introduce the ratio $\frac{r}{R} = M$, so $r = MR$ (c)

where M is less than 1.

Then the displacement of the point P is

$$f = \frac{R}{2} \left[(1 + M) - (\operatorname{Cos} \beta + M \operatorname{Cos} 2\beta)\right] \tag{2}$$

The displacement is shown on the chart 7 in the dimensionless form, as

$\frac{f}{R}$; only one half of the cycle is shown, because the second half is a mirror image of the first. Consider as one cycle when the crank with the radius R makes one revolution or one cycle is made during $\frac{1}{N}$ sec. Again, it is considered that both cranks rotate uniformly, so their angles of rotation are proportional to time.

The variable time $t = \frac{\beta}{2\pi} \cdot \frac{1}{N}$ sec.

so $\beta = 2\pi Nt$ (d)

During the differentiation to find the velocity and the acceleration equations, β should be in radians (expression d), but in final form of equation β can be used in degrees.

The velocity equation

$$v = \frac{df}{dt} = \pi NR\ (\text{Sin}\ \beta + 2M\ \text{Sin}\ 2\beta) \qquad (3)$$

The chart 8 shows values of

$$\frac{v}{RN} = \pi\ (\text{Sin}\ \beta + 2M\ \text{Sin}\ 2\beta) \qquad (3_1)$$

The acceleration equation

$$a = \frac{dv}{dt} = 2\pi^2 N^2 R\ (\text{Cos}\ \beta + 4M\ \text{Cos}\ 2\beta) \qquad (4)$$

The chart 9 presents

$$\frac{a}{RN^2} \qquad (4_1)$$

All three charts show variation of $\frac{f}{R}$, $\frac{v}{RN}$ and $\frac{a}{RN^2}$ for β varying from 0° to 180°; the other half of diagrams is the mirror image.

Some limit values for the displacement can be found analytically; they happen when the velocity is zero.

The velocity equation can be written as

$$v = \pi RN\ \text{Sin}\ \beta\ (1 + 4M\ \text{Cos}\ \beta) \qquad (3_2)$$

The velocity can be zero when either Sin β or $(1 + 4M\ \text{Cos}\ \beta)$ are zero.

(1) Sin $\beta = 0$ under two conditions:

(a) $\beta = 0°$; then Cos β and Cos $2\beta = 1$.

Introduce these values in the equation (2) for the displacement

$$f = \frac{R}{2}\left[(1 + M) - (1 + M)\right] = 0.$$

(b) $\beta = 180°$; Cos $\beta = -1$; Cos $2\beta = 1$; then

$$f = \frac{R}{2}\left[(1 + M) - (-1 + M)\right] = 2\frac{R}{2} = R.$$

It means at the middle of the cycle the displacement is independent of the ratio M.

(2) $(1 + 4M \cos\beta) = 0$

$$\cos\beta = -\frac{1}{4M} \text{ and } \cos 2\beta = 2\cos^2\beta - 1 = \frac{1}{8M^2} - 1;$$

$$f = \frac{R}{2}\left[(1 + M) - \left(-\frac{1}{4M} + \frac{M}{8M^2} - M\right)\right] =$$

$$= \frac{R}{2}\left(1 + 2M + \frac{1}{8M}\right) = R\left(0.5 + M + \frac{1}{16M}\right)$$

When M is sufficiently large, f may reach the magnitude above R, but if M is below a certain value, the maximum value of f is R. This value can be determined for the conditions when

$$M + \frac{1}{16M} = \frac{1}{2}$$

The solution is $M = 1/4$.

From the chart of displacements, it can be seen that when M is slightly more than $1/4$, there is practically a dwell when the point P hovers near $f - R$, as on the diagram for $M = 0.3$.

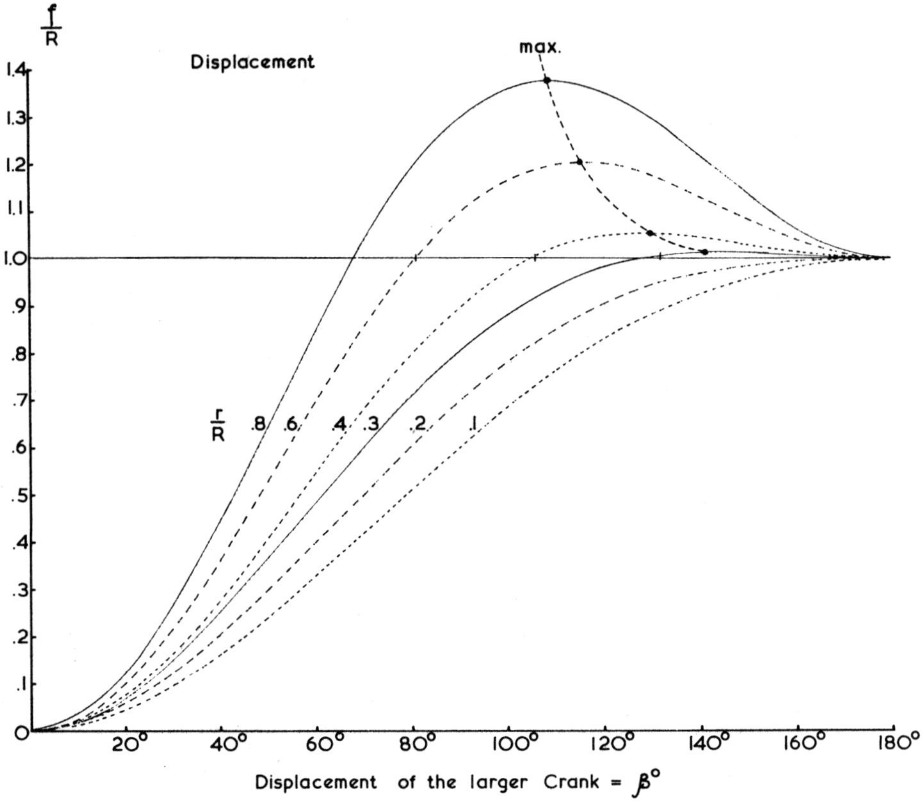

Chart 7. Double Crank Mechanism

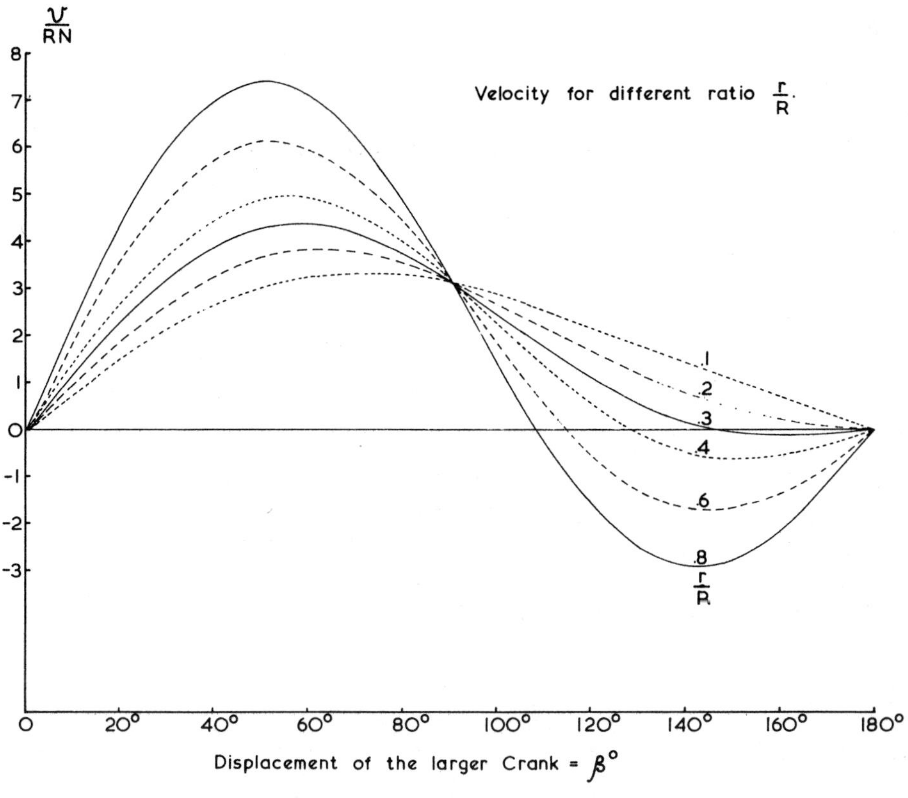

Chart 8. Double Crank Mechanism

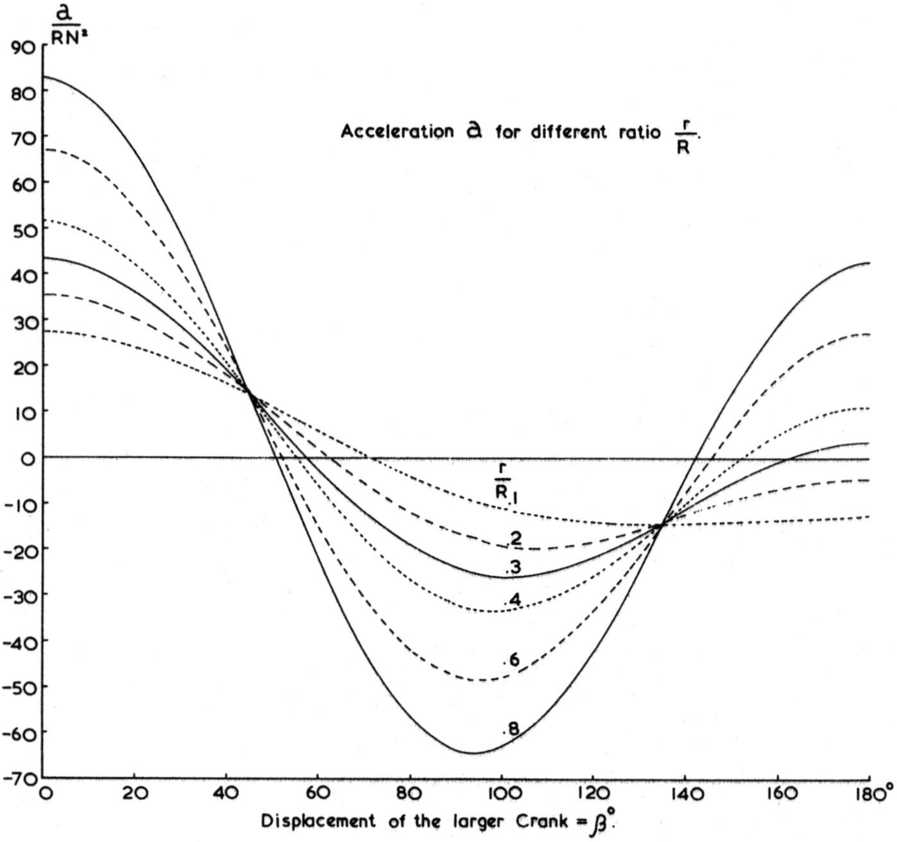

Chart 9. Double Crank Mechanism

IIa. Elliptical Gears

The analysis was made after an inspection of an experimental machine which could not be operated even at half of the speed predicted by designers; vibrations dislodged products handled by the machine, so numerous jams decreased the efficiency of the machine and the amount of waste was not acceptable. Four sets of elliptical gears, used in the machine, were the only unusual detail of the machine.

The results of the analysis proved that elliptical gears could be the main reason of the failure. Nevertheless, the analysis is presented as an example of more complicated problems.

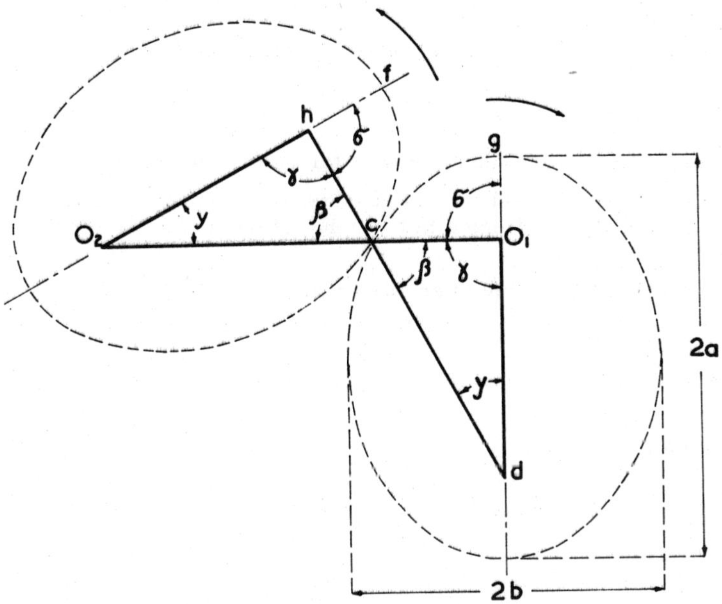

Fig. II - 7

O_1 – axis of the driving gear;

O_2 – axis of the driven gear.

The initial position of gears is when points g and f are in contact. At that moment, the driven gear has the lowest velocity.

Geometric properties of the ellipse were used to draw the sketch—Fig.II-7.

$$\overline{O_1O_2} = \overline{dh} = 2a = \overline{O_1Cd} = \overline{O_2Ch} = \overline{O_1C} + \overline{Cd} \qquad \text{(a)}$$

$$e = \frac{\sqrt{a^2 - b^2}}{a} \tag{b}$$

$$\overline{O_1d} = \overline{O_2h} = 2\sqrt{a^2 - b^2} \tag{c}$$

$$\overline{O_1C} = \overline{hc} = a - ex \text{ and } \overline{O_2C} = \overline{dc} = a + ex$$

where x is measured from the center of the ellipse.

The analysis of the movement of gears is made for the moment (shown on Fig. II-7) when the driving gear rotated on the angle σ.

$$\angle gO_1C = \sigma$$

At that moment the driven gear is rotated on the angle y

$$\angle fO_2C = y$$

The problem is to express the angle y in terms of the angle σ.

As in analyses of other mechanisms, it is assumed that the driver rotates at uniform speed, so the changes in the value of the angle σ are proportional to the time t.

Study the triangle O_1CD, which permits to write

$$\frac{\overline{O_1C}}{\text{Sin } y} = \frac{\overline{cd}}{\text{Sin } \gamma} = \frac{\overline{O_1d}}{\text{Sin } \beta} \tag{d}$$

$$\overline{O_1C} = \overline{cd} \cdot \frac{\text{Sin } y}{\text{Sin } \gamma};$$

introduce it into the expression (a)

$$\overline{cd} \cdot \frac{\text{Sin } y}{\text{Sin } \gamma} + \overline{cd} = 2a, \text{ therefore } \overline{cd} = 2a \frac{\text{Sin } \gamma}{\text{Sin } \gamma + \text{Sin } y} \tag{e}$$

from eq. (d);

$$\overline{cd} = \overline{O_1d} \frac{\text{Sin } \gamma}{\text{Sin } \beta},$$

which represents the same side of the triangle.

Replace O_1d by the expression (c) and combine with the expression (e)

$$2\sqrt{a^2 - b^2} \frac{\text{Sin } \gamma}{\text{Sin } \beta} = 2a \frac{\text{Sin } \gamma}{\text{Sin } \gamma + \text{Sin } y}$$

Introduce e from the expression (b), then

$$e = \frac{\text{Sin } \beta}{\text{Sin } \gamma + \text{Sin y}} \qquad (1)$$

Replace angles β and γ

$$\gamma = 180° - \sigma°; \qquad \text{Sin } \gamma = \text{Sin } \sigma \qquad (f)$$

$$\beta = 180° - \gamma° - y° = \sigma - y; \qquad \text{Sin } \beta = \text{Sin } \sigma \,.\, \text{Cos y} - \text{Cos } \sigma \,.\, \text{Sin y} \qquad (g)$$

Then the equation (1) will have only angles σ and y.

$$\text{e Sin y} + \text{e Sin } \sigma = \text{Sin } \sigma \text{ Cos y} - \text{Cos } \sigma \text{ Sin y} \qquad (1_1)$$

Replace Cos y by $\sqrt{1 - \text{Sin}^2\text{y}}$.

$$\text{Sin } \sigma \sqrt{1 - \text{Sin}^2\text{y}} = \text{e Sin y} + \text{e Sin } \sigma + \text{Cos } \sigma \text{ Sin}^2 \sigma$$

$$\text{Sin}^2\sigma - \text{Sin}^2\sigma \text{ Sin}^2\text{y} = \text{Sin}^2\text{y}(\text{e}^2 + 2\text{e Cos } \sigma + \text{Cos}^2\sigma) +$$

$$+ \text{Sin y} \,.\, 2\text{e sin } \sigma(\text{e} + \text{Cos } \sigma) + \text{e}^2 \text{ Sin}^2 \sigma$$

$$\text{Sin}^2\text{y}(1 + \text{e}^2 + 2\text{e Cos } \sigma) + \text{Sin y}(2\text{e}^2\text{Sin } \sigma + 2\text{e Sin } \sigma \text{ Cos } \sigma) -$$

$$- (\text{Sin}^2\sigma - \text{e}^2\text{Sin}^2\sigma) = 0$$

$$\text{Sin y} = \frac{-(\text{e}^2\text{Sin } \sigma + \text{e Sin } \sigma \text{ Cos } \sigma) \pm \sqrt{(\text{e}^2\text{Sin } \sigma + \text{e Sin } \sigma \text{ Cos } \sigma)^2 + (\text{Sin}^2\sigma - \text{e}^2\text{Sin}^2\sigma)(1 + \text{e}^2 + 2\text{e Cos } \sigma)}}{1 + \text{e}^2 + 2\text{e Cos } \sigma}$$

The numerator under the radical sign can be transformed into $\text{Sin}^2\sigma(1 + \text{e Cos } \sigma)^2$, so

$$\text{Sin y} = \frac{-\text{e}^2\text{Sin } \sigma - \text{e Sin } \sigma \text{ Cos } \sigma \pm \text{Sin } \sigma(1 + \text{e Cos } \sigma)}{1 + \text{e}^2 + 2\text{e Cos } \sigma} \qquad (1_2)$$

The expression (1_2) has two solutions with signs – and +.

With – sign

$$\text{Sin y}_1 = -\text{Sin } \sigma$$

With + sign

$$\text{Sin y}_2 = \text{Sin } \sigma \frac{1 - \text{e}^2}{1 + \text{e}^2 + 2\text{c Cos } \sigma}$$

It is evident that the + sign should be used, so

$$\text{Sin } y = \text{Sin } \sigma \frac{1 - e^2}{1 + e^2 + 2e \text{ Cos } \sigma} \tag{2}$$

The angular displacement of the driven gear is

$$s = \text{arc Sin}\left[\text{Sin } \sigma \frac{1 - e^2}{1 + e^2 + 2e \text{ Cos } \sigma}\right] \tag{2_1}$$

On the chart 10 the variation of the angle y is presented as a function of the angle σ for the first 180°.

To find equations for velocity and for acceleration, it is necessary to express the variable angle σ in radians and as a function of the time t. The angular speed of the driver is $2\pi N$, where N is in revolutions per second.

Then $\sigma = 2\pi Nt$

The angular velocity

$$w = \frac{ds}{dt} = \frac{dx/dt}{\sqrt{1 - x^2}}, \text{ where } x = \text{arc Sin}\left(\frac{\text{Sin } 2\pi Nt \,.\, (1 - e^2)}{1 + e^2 + 2e \text{ Cos } 2\pi Nt}\right)$$

When the differentiation is completed, use σ instead of $2\pi Nt$

$$\frac{dx}{dt} = 2\pi N(1 - e^2)$$

$$\frac{1}{\sqrt{1 - x^2}} = \frac{1 + e^2 + 2e \text{ Cos } \sigma}{\sqrt{(1 + e^2 + 2e \text{ Cos } \sigma)^2 - (1 - e^2) \text{ Sin }^2\sigma}}$$

The expression under the radical can be transformed into $[2e + (1 + e^2) \text{ Cos } \sigma]^2$, so

$$\frac{1}{\sqrt{1 - x^2}} = \frac{1 + e^2 + 2e \text{ Cos } \sigma}{2e + (1 + e^2) \text{ Cos } \sigma}$$

and $$w = 2\pi N \frac{1 - e^2}{1 + e^2 + 2e \text{ Cos } \sigma} \tag{3}$$

On the chart 11, it is presented as

$$\frac{w}{N} = 2\pi \frac{1 - e^2}{1 + e^2 + 2e \text{ Cos } \sigma} \tag{3_1}$$

The angular acceleration

$$\Omega = \frac{dw}{dt} = (2\pi N)^2\, 2e(1 - e^2)\, \frac{\mathrm{Sin}\,\sigma}{(1 + e^2 + 2e\,\mathrm{Cos}\,\sigma)^2} \tag{4}$$

On the chart 12, it is presented as

$$\frac{\Omega}{N^2} = (2\pi)^2\, \frac{2e(1 - e^2)\,\mathrm{Sin}\,\sigma}{(1 + e^2 + 2e\,\mathrm{Cos}\,\sigma)^2} \tag{4_1}$$

The most important property of the elliptical gear is the variation of velocity, so for the designer it is necessary to know either the maximum velocity or a ratio between maximum and minimum. Both limits of velocity happen when the angular acceleration $\Omega = 0$ or when

$$\frac{2e(1 - e^2)\,\mathrm{Sin}\,\sigma}{(1 + e^2 + 2e\,\mathrm{Cos}\,\sigma)^2} = 0.$$

Since the denominator cannot be infinitely large, it means the numerator should be equal zero.

e^2 is smaller than unity, therefore, $(1 - e^2)$ cannot be zero, so only Sin σ can reach 0.

It happens when $\sigma = 0°$ and $180°$; at these moments Cos $\sigma = \pm 1$.

$$w_{min} = 2\pi N\, \frac{1 - e^2}{1 + e^2 + 2e} = 2\pi N\, \frac{(1 + e)(1 - e)}{(1 + e)^2} = 2\pi N\, \frac{1 - e}{1 + e} \tag{5_1}$$

$$w_{max} = 2\pi N\, \frac{1 - e^2}{1 + e^2 - 2e} = 2\pi N\, \frac{(1 + e)(1 - e)}{(1 - e)^2} = 2\pi N\, \frac{1 + e}{1 - e} \tag{5_2}$$

The ratio $$\frac{w_{max}}{w_{min}} = \left(\frac{1 + e}{1 - e}\right)^2 \tag{5_3}$$

The charts are made for σ varying between 0° and 180°, because the other half of the cycle is the mirror image.

The defect of this mechanism is in the fact that maximum acceleration happens close to $\sigma = 180°$ and it changes into maximum deceleration quickly, if $e \geqslant 0.4$.

The advantage of elliptical gears is in a high ratio $\frac{\text{max. velocity}}{\text{min. velocity}}$.

It should be remembered that the analysis is made for ideal elliptical gears, when pitch lines of gears have the form of an ellipse; any deviation from an ellipse may increase dynamic forces and/or make the change of forces more violent.

The table shows the connections between the ratio $\frac{a}{b}$ and the values of e and $\frac{\text{max. velocity}}{\text{min. velocity}}$.

a/b	1.01	1.02	1.03	1.05	1.1	1.2
e	0.1404	0.197	0.24	0.3049	0.4166	0.5528
$\frac{V_{max}}{V_{min}}$	1.76	2.22	2.66	3.52	5.93	12.05

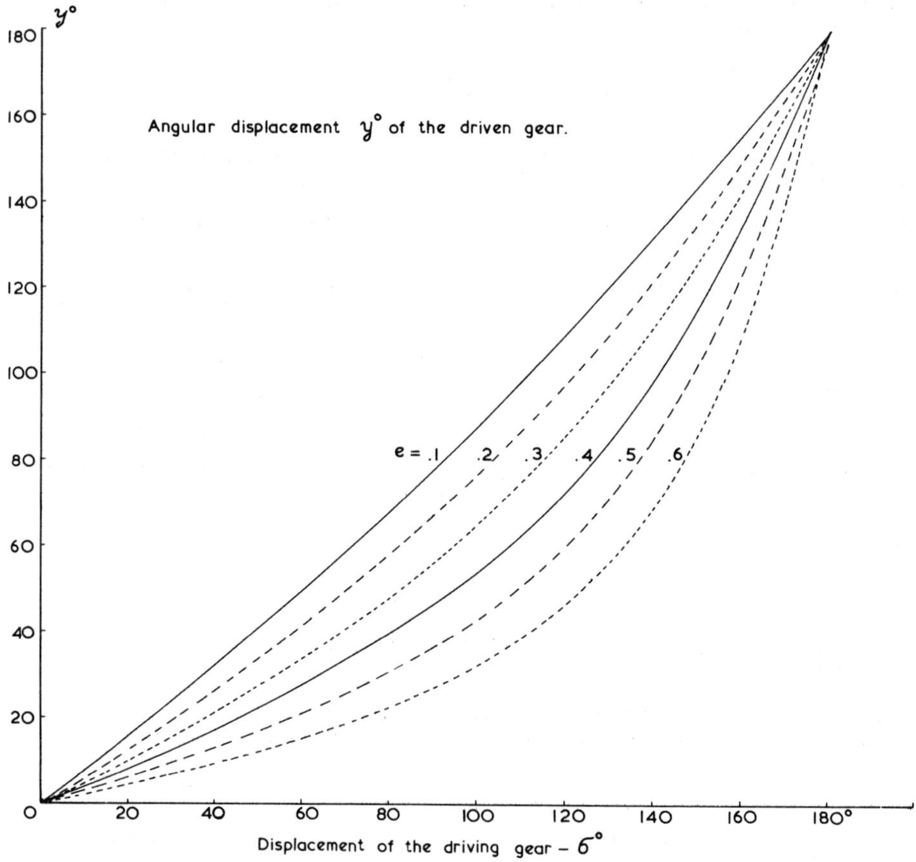

Chart 10. **Elliptical Gears**

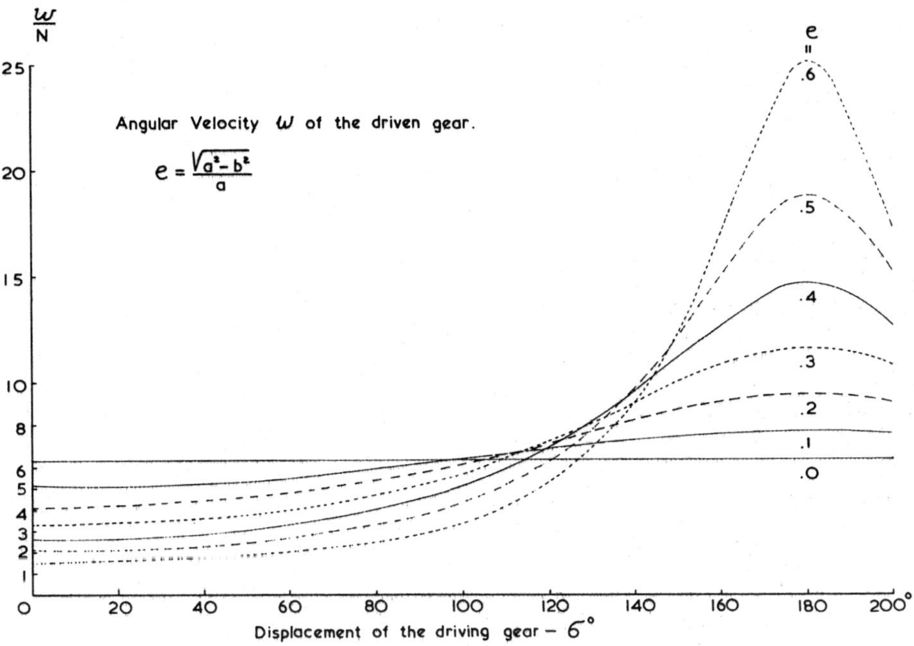

Chart 11. Elliptical Gears

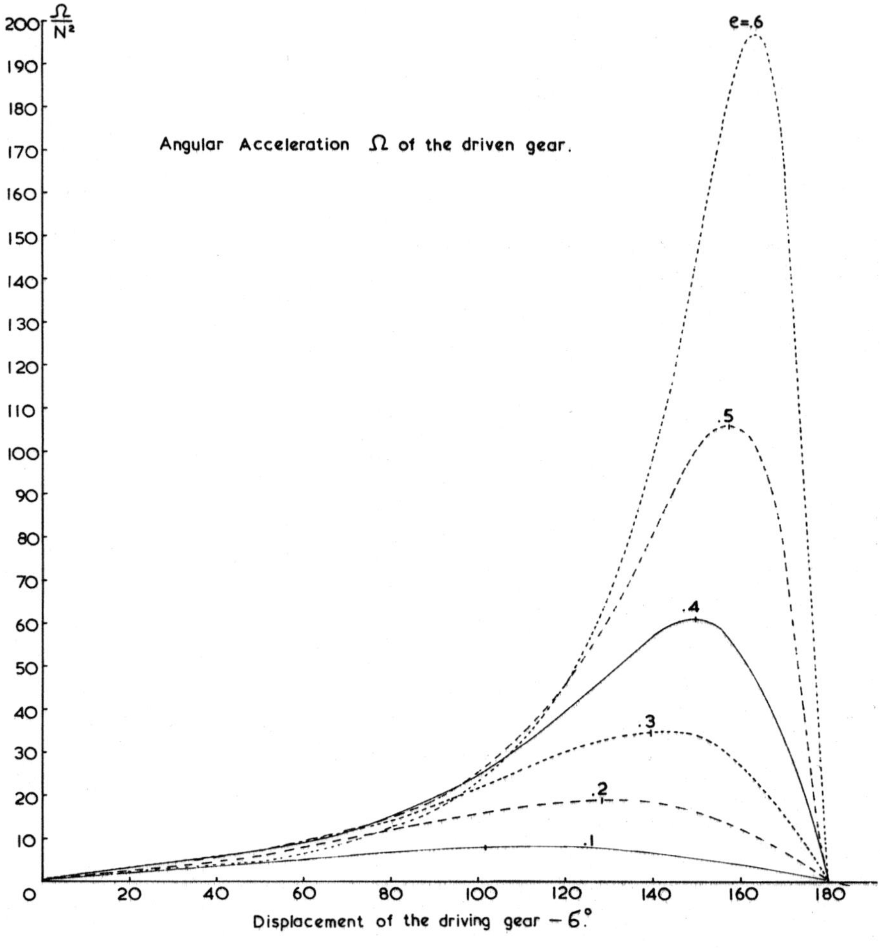

Chart 12. Elliptical Gears

IIb. Offset Cranks or Slot and Crank

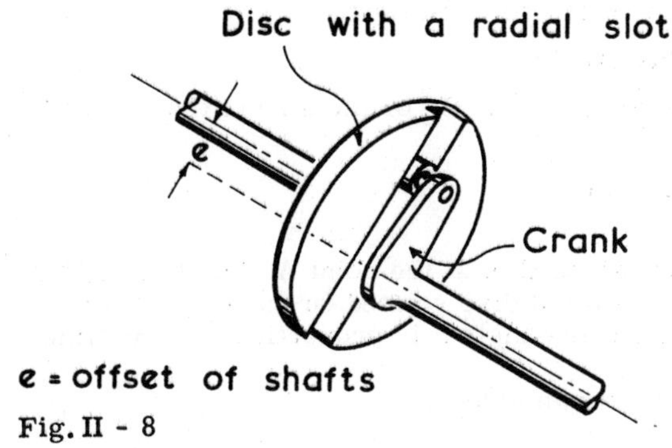

Fig. II - 8

Analyses were made using either shaft as a driver. It was found that when the crank is the driver, the acceleration diagram is of the same type as in the case of the elliptical gears, which is considered undesirable due to a fast change from + max. to – max. acceleration.

The mechanism has better dynamic characteristics when the slotted disc is used as the driver, the + and – maxima are separated wider, as it can be seen on the chart 15.

A test was made using either shaft as the driver; the results confirmed the findings of the analyses.

Only the analysis of the case, when the disc is the driver, is presented.

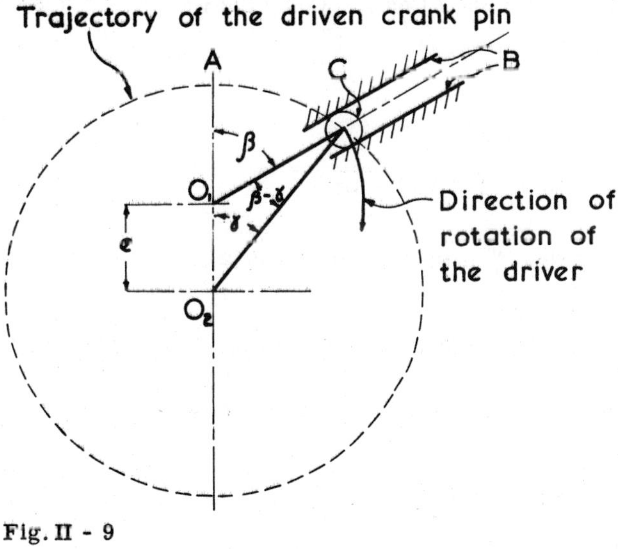

Fig. II - 9

O_1 — axis of the driving shaft which carries the slotted disc;

B — sides of the slot; only a part of the slot is shown;

O_2 — axis of the driven shaft with the crank;

R — O_2C—radius of the crank;

e — distance between axes O_1 and O_2 or the size of the "offset" of shafts;

C — center of the crank pin.

Select as the initial position when C is at the point A. For the analysis, select the moment when the slotted disc rotated on the angle β. Due to uniform rotation of the driver, the angle β is proportional to the time t.

The triangle O_1O_2C has two known sides

$$\overline{O_1O_2} = e \text{ and } \overline{O_2C} = R.$$

For this triangle

$$\frac{e}{\text{Sin }(\beta - \gamma)} = \frac{R}{\text{Sin }(180 - \beta)} = \frac{R}{\text{Sin }\beta} \qquad \text{(a)}$$

The problem is to express the angle γ in terms of the angle β

From the equation (a)

$$\frac{e}{R}\text{ Sin }\beta = \text{Sin }(\beta - \gamma) = \text{Sin }\beta\text{ Cos }\gamma - \text{Cos }\beta\text{ Sin }\gamma$$

Introduce

$$\text{Cos }\gamma = \sqrt{1 - \text{Sin}^2\,\gamma}$$

$$\frac{e}{R}\text{ Sin }\beta + \text{Cos }\beta\text{ Sin }\gamma = \text{Sin }\beta\sqrt{1 - \text{Sin}^2\,\gamma}$$

$$\left(\frac{e}{R}\text{ Sin }\beta + \text{Cos }\beta\text{ Sin }\gamma\right)^2 = \text{Sin}^2\,\beta - \text{Sin}^2\,\beta\text{ Sin}^2\,\gamma$$

$$\text{Sin}^2\,\gamma\,(\text{Cos}^2\,\beta + \text{Sin}^2\,\beta) + \text{Sin }\gamma\,.\,2\frac{e}{R}\text{ Sin }\beta\text{ Cos }\beta - \left[1 - \left(\frac{e}{R}\right)^2\right]\text{Sin}^2\,\beta = 0$$

$$\text{Sin }\gamma = -\frac{e}{R}\text{ Sin }\beta\text{ Cos }\beta \pm \text{Sin }\beta\sqrt{1 - \left(\frac{e}{R}\right)^2(1 - \text{Cos}^2\,\beta)} =$$

$$= -\frac{e}{R}\,.\,\frac{\text{Sin }2\,\beta}{2} \pm \text{Sin }\beta\sqrt{1 - \left(\frac{e}{R}\right)^2\text{Sin}^2\,\beta} \qquad (1)$$

The chart 13 shows γ as a function of the angle β for one half of the revolution of the driver.

When the angle β changes from 0° to 90°, Sin β and Sin 2β are positive, and also Sin γ is positive.

Therefore, the + sign should be used

$$\operatorname{Sin}\gamma = \operatorname{Sin}\beta\sqrt{1-\left(\frac{e}{R}\right)^2\operatorname{Sin}^2\beta} - \frac{e}{R}\cdot\frac{\operatorname{Sin}2\beta}{2} \tag{2}$$

The angular displacement is

$$s = \operatorname{arc}\operatorname{Sin}\left(\operatorname{Sin}\beta\sqrt{1-\left(\frac{e}{R}\right)^2\operatorname{Sin}^2\beta} - \frac{e}{R}\cdot\frac{\operatorname{Sin}2\beta}{2}\right) \tag{2_1}$$

The variable angle $\beta = 2\pi Nt$.

The angular velocity

$$w = \frac{ds}{dt} = \frac{\frac{dx}{dt}}{\sqrt{1-x^2}}, \text{ where } x = \operatorname{Sin}\beta\sqrt{1-\left(\frac{e}{R}\right)^2\operatorname{Sin}^2\beta} - \frac{e}{R}\operatorname{Sin}\beta\operatorname{Cos}\beta.$$

$$\frac{dx}{dt} = 2\pi N\left[\frac{-\frac{e}{R}^2\operatorname{Sin}^2\beta\operatorname{Cos}\beta}{\sqrt{1-\left(\frac{e}{R}\right)^2\operatorname{Sin}^2\beta}} + \operatorname{Cos}\beta\sqrt{1-\left(\frac{e}{R}\right)^2\operatorname{Sin}^2\beta} + \right.$$

$$\left. + \frac{e}{R}\operatorname{Sin}^2\beta - \frac{e}{R}\operatorname{Cos}^2\beta\ ;\right]$$

$$\sqrt{1-x^2} = \operatorname{Cos}\beta\sqrt{1-\left(\frac{e}{R}\right)^2\operatorname{Sin}^2\beta} + \frac{e}{R}\operatorname{Sin}^2\beta.$$

After making necessary transformation, the equation will have the form

$$w = 2\pi N\left(1-\frac{\frac{e}{R}\operatorname{Cos}\beta}{\sqrt{1-\left(\frac{e}{R}\right)^2\operatorname{Sin}^2\beta}}\right) \tag{3}$$

The chart 14 shows

$$\frac{w}{N} = 2\pi\left(1-\frac{\frac{e}{R}\operatorname{Cos}\beta}{\sqrt{1-\left(\frac{e}{R}\right)^2\operatorname{Sin}^2\beta}}\right) \tag{3_1}$$

The angular acceleration

$$\Omega = \frac{dw}{dt} = \frac{d}{dt}\left(-2\pi N \frac{\frac{e}{R}\,\mathrm{Cos}\,\beta}{\sqrt{1-\left(\frac{e}{R}\right)^2 \mathrm{Sin}^2\,\beta}}\right) =$$

$$= -2\pi N\frac{e}{R}\,\frac{-2\pi N\,\mathrm{Sin}\,\beta\sqrt{1-\left(\frac{e}{R}\right)^2 \mathrm{Sin}^2\,\beta} - \mathrm{Cos}\,\beta\,\frac{-\left(\frac{e}{R}\right)^2 2\,\mathrm{Sin}\,\beta\,\mathrm{Cos}\,\beta\,.\,2\pi N}{2\sqrt{1-\left(\frac{e}{R}\right)^2 \mathrm{Sin}^2\,\beta}}}{1-\left(\frac{e}{R}\right)^2 \mathrm{Sin}^2\,\beta} =$$

$$= (2\pi N)^2\,\frac{\frac{e}{R}\left[1-\left(\frac{e}{R}\right)^2\right]\mathrm{Sin}\,\beta}{\left(\sqrt{1-\left(\frac{e}{R}\right)^2 \mathrm{Sin}^2\,\beta}\right)^3} \qquad (4)$$

The chart 15 presents

$$\frac{\Omega}{N^2}$$

The mechanism is used to produce varying velocity during one revolution of the driver; therefore, it is necessary to determine the maximum and the minimum velocity. It happens when $\Omega = 0$.

In the equation of the acceleration (equation 4) the expression

$$1-\left(\frac{e}{R}\right)^2 \mathrm{Sin}^2\,\beta$$

cannot be infinity and

$$1-\left(\frac{e}{R}\right)^2$$

cannot be equal zero because when $e = R$, the mechanism is not operative.

So, $\Omega = 0$ only when $\mathrm{Sin}\,\beta = 0$ or when $\beta = 0°$ or $180°$, then $\mathrm{Cos}\,\beta = \pm 1$.

Introduce it into equation (3)

$$\text{When } \beta = 0°;\ w = 2\pi N\left(1-\frac{\frac{e}{R}\,.\,1}{\sqrt{1}}\right) = 2\pi N\left(1-\frac{e}{R}\right) = w_{min} \qquad (5_1)$$

When $\beta = 180°$; $w = 2\pi N\left(1 - \frac{\frac{e}{R} . (-1)}{\sqrt{1}}\right) = 2\pi N\left(1 + \frac{e}{R}\right) = w_{max}$ (5_2)

The ratio $\frac{w_{max}}{w_{min}} = \frac{R + e}{R - e}$ (5_3)

The table shows the influence of the ratio $\frac{e}{R}$ on ratios $\frac{w_{max}}{w_{min}}$;

$\frac{e}{R}$	0.2	1/3	1/2	2/3	3/4	0.9	0.95	0.98
$\frac{w_{max}}{w_{min}}$	1.5	2	3	5	7	19	39	99

This mechanism is effective to slow down the driven shaft, so it can be used for an approximate intermittent motion.

Maximum or minimum acceleration is at the moments when the derivative of Ω is zero.

$$\frac{d\Omega}{dt} = \text{Cos}\ \beta\left[1 + 2\left(\frac{e}{R}\right)^2 \text{Sin}^2\ \beta\right] = 0$$

That expression is satisfied when Cos $\beta = 0$ and $\beta = 90°$ and $270°$

When $\beta = 90°$; $\Omega = (2\pi N)^2 \frac{\frac{e}{R}}{\sqrt{1 - \left(\frac{e}{R}\right)^2}}$;

When $\beta = 270°$; $\Omega = -(2\pi N)^2 \frac{\frac{e}{R}}{\sqrt{1 - \left(\frac{e}{R}\right)^2}}$;

Therefore

$$\Omega_{max} = \pm(2\pi N)^2 \frac{\frac{e}{R}}{\sqrt{1 - \left(\frac{e}{R}\right)^2}}. \qquad \text{(b)}$$

The advantage of this mechanism is in possible changes of the ratio $\frac{e}{R}$ without stopping the machine, so the velocities can be adjusted easily.

Also, the variation of the acceleration is more advantageous than in the case of elliptical gears: the maximum acceleration happens when $\beta = 90°$ and the maximum deceleration—when $\beta = 270°$, so the change from + maximum to − maximum happens during a longer period. In actual practice, this mechanism was used successfully at 140 RPM, but during experiments, the movements were smooth even at 300 RPM.

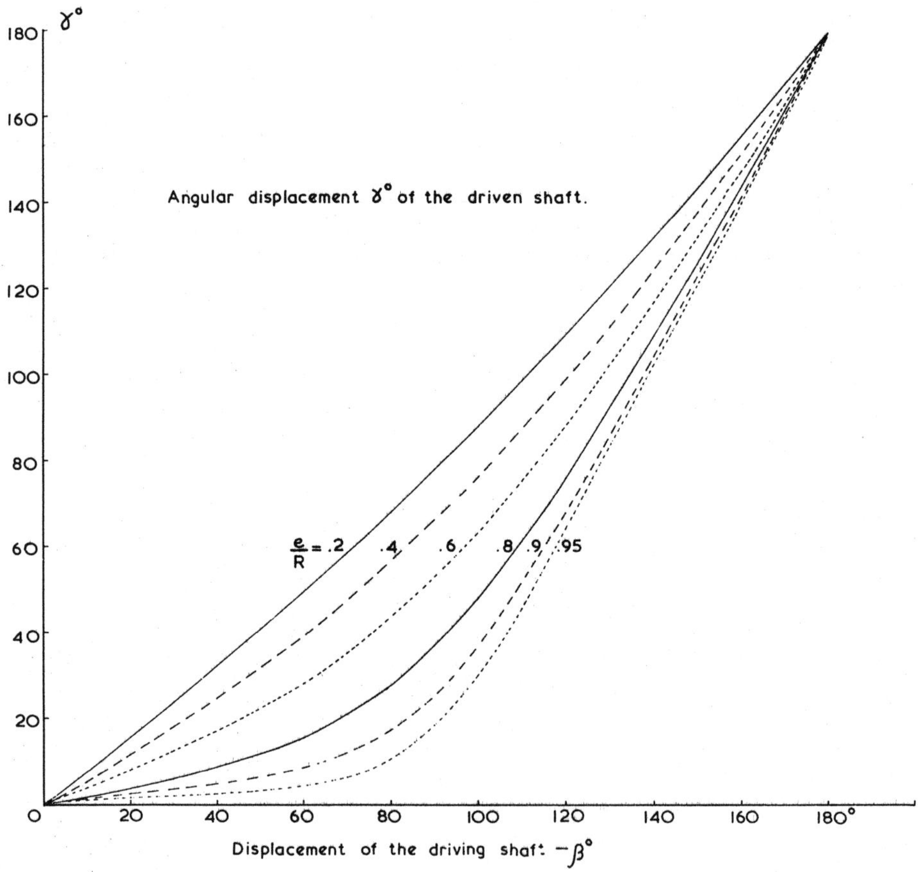

Chart 13. Offset Cranks

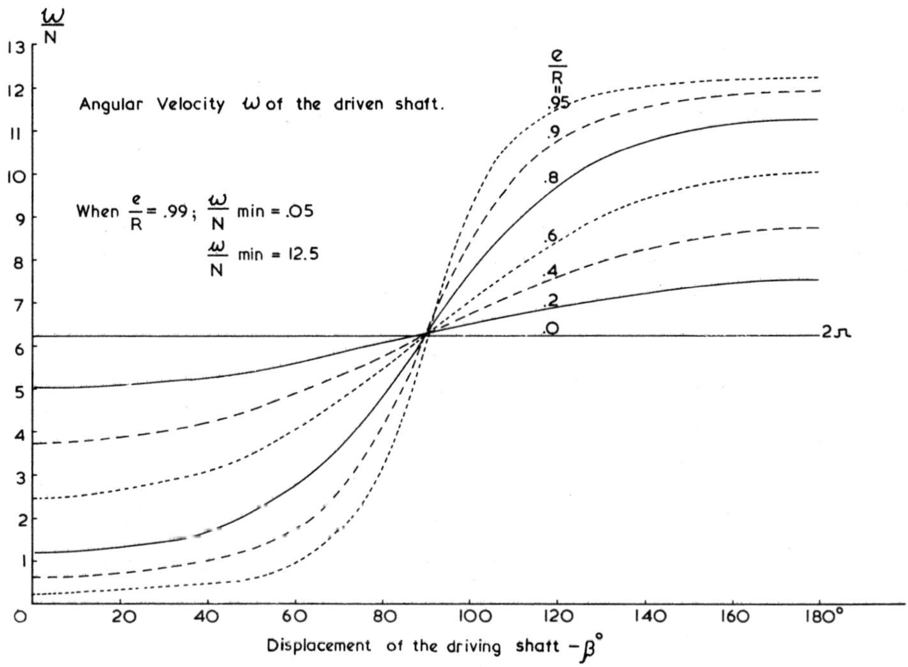

Chart 14. Offset Cranks

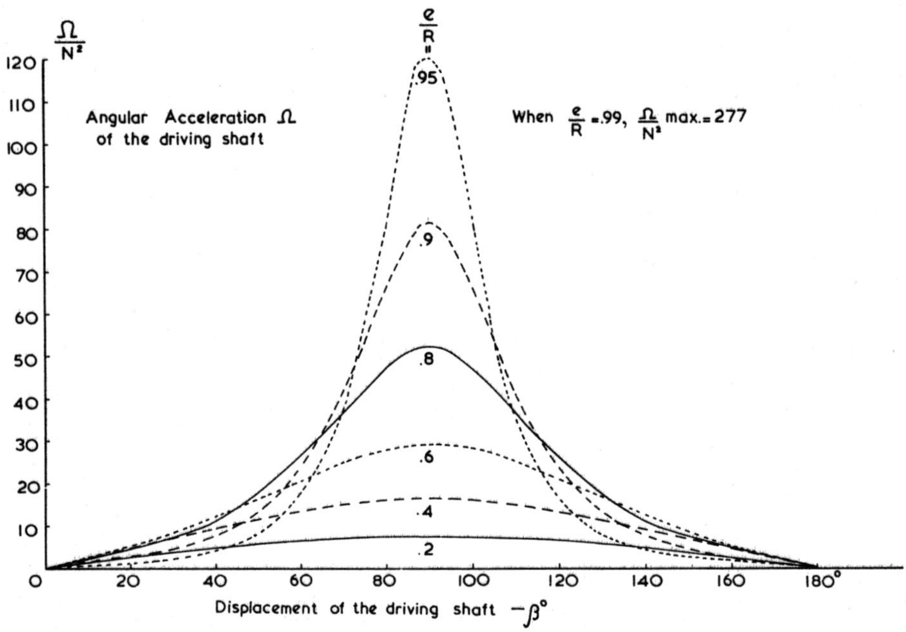

Chart 15. Offset Cranks

IIIa. Geneva Drive-external

Driven wheel—can have 3 or more slots

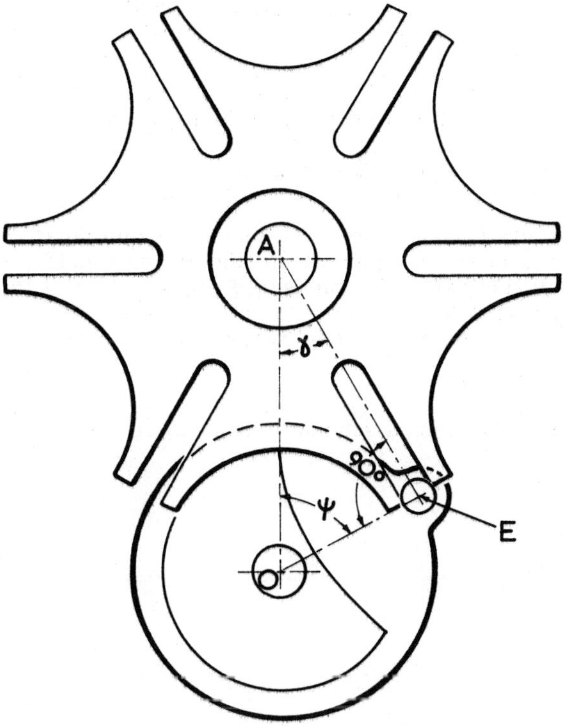

Fig. II -10

The initial position for the analysis is shown on the drawing: The driving roller enters the slot and the angle OEA = 90°.

- O — axis of the driver;
- A — axis of the driven wheel;
- h — number of slots in the wheel;
- 2ψ — indexing period of the driver;
- 2γ — rotation of the wheel during the indexing period.

Properties of the mechanism:

$$\gamma = \frac{180^\circ}{n}; \qquad \psi = 90^\circ - \gamma^\circ = 90\,\frac{n-2}{n}.$$

The locking of the wheel during dwells is performed by sliding circular surfaces. The maximum locking angle on the wheel $= 2\psi$

The maximum locking angle on the driver:

(a) in the case of the driver with one roller

$$360° - 2\psi = 180 \frac{n + 2}{n}$$

(b) in case of the driver with two driving rollers, 180° apart

$$180° - \psi° = 90 \frac{n + 2}{n}, \text{ or twice smaller.}$$

The driver with two rollers produces two strokes per revolution; to produce a certain number of strokes per minute, the driver should rotate at a half speed. The indexing period 2ψ, measured on the driver is the same, but it occupies a part of 180°, while in the case of one roller, it is a part of 360°. Therefore, in the case of two rollers, the ratio $\frac{\text{indexing period}}{\text{time of one cycle}}$ is twice higher.

For the analysis of the mechanism, a schematic drawing is used (Fig. II-11).

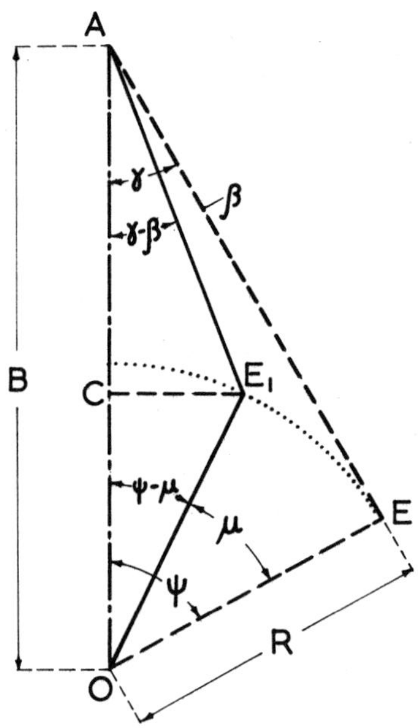

Fig. II - 11

OE — represents the initial position of the radius to the center E of the driving roller;

OE_1 — represents the analyzed position of the same radius after it rotated on the angle μ;

R — the length of the radius OE (OE_1);

B — center to center distances between the shaft centers O and A.

The problem is to express the angle β in terms of the angle μ.

From the triangle OCE,

$$\overline{CE_1} = R \text{ Sin } (\psi - \mu) \quad \text{(a)}$$

$$\overline{OC} = R \text{ Cos } (\psi - \mu) \quad \text{(b)}$$

In the triangle ACE_1

$$\overline{AC} = B - \overline{OC} = B - R \text{ Cos } (\psi - \mu) \quad \text{(c)}$$

$$\tan (\gamma - \beta) = \frac{\overline{CE_1}}{\overline{AC}} = \frac{R \text{ Sin } (\psi - \mu)}{B - R \text{ Cos } (\psi - \mu)} \quad (1)$$

Separate trigonometric functions of the angles ψ, γ, μ and β

$$\text{Sin } (\psi - \mu) = \text{Sin } \psi \text{ Cos } \mu - \text{Cos } \psi \text{ Sin } \mu \quad (a_1)$$

$$\text{Cos } (\psi - \mu) = \text{Cos } \psi \text{ Cos } \mu + \text{Sin } \psi \text{ Sin } \mu \quad (b_1)$$

$$\tan (\gamma - \beta) = \frac{\tan \gamma - \tan \beta}{1 + \tan \gamma \, . \, \tan \beta} \quad \text{(d)}$$

From the triangle OAE

$$R = B \text{ Sin } \gamma \text{ or } B \, . \, \tan \gamma = \frac{R}{\text{Cos } \gamma} \text{ or } B = \frac{R}{\text{Sin } \gamma} \quad \text{(e)}$$

From the properties of the mechanism $\psi = 90 - \gamma$; therefore,

$$\text{Sin } \psi = \text{Cos } \gamma \text{ and Cos } \psi = \text{Sin } \gamma \quad \text{(f)}$$

$$\text{Sin } (\psi - \mu) = \text{Cos } \gamma \text{ Cos } \mu - \text{Sin } \psi \text{ Sin } \mu \quad (a_2)$$

$$\text{Cos } (\psi - \mu) = \text{Sin } \gamma \, . \, \text{Cos } \mu + \text{Cos } \gamma \, . \, \text{Sin } \mu \quad (b_2)$$

Introduce expressions (a_2), (b_2), (d) and (e) into the equation (1)

$$\tan \gamma - \tan \beta = \frac{R \text{ Sin } (\psi - \mu)}{\frac{R}{\text{Sin } \gamma} - R \text{ Cos } (\psi - \mu)} \, . \, (1 + \tan \gamma \, . \, \tan \beta)$$

After transformations

$$\tan\beta\,(1 - \operatorname{Sin}\gamma\,.\,\operatorname{Cos}\gamma\,.\,\operatorname{Sin}\mu - \tan\gamma\,.\,\operatorname{Sin}^2\gamma\,.\,\operatorname{Sin}\mu) =$$

$$= \tan\gamma - \tan\gamma\,.\,\operatorname{Sin}^2\gamma\,.\,\operatorname{Cos}\mu - \operatorname{Sin}\gamma\,.\,\operatorname{Cos}\gamma\,.\,\operatorname{Cos}\mu$$

$$\tan\beta = \frac{\tan\gamma(1 - \operatorname{Sin}^2\gamma\operatorname{Cos}\mu - \operatorname{Cos}^2\gamma\,.\,\operatorname{Cos}\mu)}{1 - \operatorname{Sin}\gamma\,.\,\operatorname{Cos}\gamma\operatorname{Sin}\mu - \tan\gamma\,.\,\operatorname{Sin}^2\gamma\operatorname{Sin}\mu} =$$

$$= \frac{\tan\gamma\,(1 - \operatorname{Cos}\mu)}{\tan\gamma\left(\dfrac{1}{\tan\gamma} - \operatorname{Cos}^2\gamma\operatorname{Sin}\mu - \operatorname{Sin}^2\gamma\operatorname{Sin}\mu\right)}$$

$$\tan\beta = \frac{1 - \operatorname{Cos}\mu}{\dfrac{1}{\tan\gamma} - \operatorname{Sin}\mu} \qquad (2)$$

The angular displacement of the wheel is

$$s = \operatorname{arc}\tan\left(\frac{1 - \operatorname{Cos}\mu}{\dfrac{1}{\tan\gamma} - \operatorname{Sin}\mu}\right) \text{ in radians}$$

On the chart 16, the displacement of the wheel for different number n of slots in the wheel is presented in the form of the ratio $\frac{\beta}{2\gamma}$, while the base of the chart represents the angle μ. The chart for displacement can be used also for lineal displacement of the driven wheel.

It is assumed that the driver rotates uniformly at N rps, then the variable angle μ is

$$\mu = 2\pi Nt \qquad (g)$$

During the differentiation, μ is replaced by equation (g), but in final expression for the angular velocity and the acceleration, it is restored.

The angular velocity

$$w = \frac{ds}{dt} = \frac{dx/dt}{1 + x^2}, \quad \text{where} \quad x = \frac{1 - \operatorname{Cos}\mu}{\dfrac{1}{\tan\gamma} - \operatorname{Sin}\mu}$$

$$\frac{dx}{dt} = 2\pi N\,\frac{\dfrac{\operatorname{Sin}\mu}{\tan\gamma} + \operatorname{Cos}\mu - 1}{\left(\dfrac{1}{\tan\gamma} - \operatorname{Sin}\mu\right)^2};$$

$$1 + x^2 = 1 + \frac{(1 - \text{Cos}\ \mu)^2}{\left(\frac{1}{\tan \gamma} - \text{Sin}\ \mu\right)^2} = \frac{\frac{1}{\tan^2 \gamma} + 2\left(1 - \frac{\text{Sin}\ \mu}{\tan \gamma} - \text{Cos}\ \mu\right)}{\left(\frac{1}{\tan \gamma} - \text{Sin}\ \mu\right)^2}$$

$$w = \pi N \frac{\frac{\text{Sin}\ \mu}{\tan \gamma} + \text{Cos}\ \mu - 1}{\frac{1}{2\tan^2 \gamma} - \left(\frac{\text{Sin}\ \mu}{\tan \gamma} + \text{Cos}\ \mu - 1\right)} \qquad (3)$$

Angular acceleration

$$\Omega = \frac{dw}{dt} = 2\pi^2 N^2 \frac{\left(1 + \frac{1}{2\tan^2 \gamma} + \frac{\text{Sin}\ \mu}{\tan \gamma} - \text{Cos}\ \mu\right)\left(\frac{\text{Cos}\ \mu}{\tan \gamma} - \text{Sin}\ \mu\right) - \left(\frac{\text{Sin}\ \mu}{\tan \gamma} + \text{Cos}\ \mu - 1\right)\left(-\frac{\text{Cos}\ \mu}{\tan \gamma} + \text{Sin}\ \mu\right)}{\left(1 + \frac{1}{2\tan^2 \gamma} - \frac{\text{Sin}\ \mu}{\tan \gamma} - \text{Cos}\ \mu\right)^2}$$

After the transformations of the numerator

$$\Omega = \frac{\pi^2 N^2}{\tan^2 \gamma} \cdot \frac{\frac{1}{\tan \gamma}\text{Cos}\ \mu - \text{Sin}\ \mu}{\left(1 + \frac{1}{2\tan^2 \gamma} - \frac{\text{Sin}\ \mu}{\tan \gamma} - \text{Cos}\ \mu\right)^2} \qquad (4)$$

Geneva drives successfully competed with cam driven mechanisms, producing intermittent motion, until the method was developed to calculate shapes of cams, based on preselected desirable acceleration diagrams. The offered analysis of Geneva drives was made when the manufacturing of cams reached a high level of precision, so they could have good dynamic characteristics of selected acceleration diagram. Therefore, the developed equations (3) and (4) were modified to allow a direct comparison with cam data.

For that purpose, it was necessary to convert the equation for angular velocity and acceleration into lineal measurements and to introduce the nomenclature developed for cams. In the equation for cams, the working period was designated as $\alpha°$ so it is equivalent to $2\psi = 180\frac{n-2}{n}$.

The time of the stroke was established as $T = \frac{\alpha°}{360\ N}$ sec. and the length of the lineal stroke was designated as F.

For the conversion from the angular to a lineal dimension, it is necessary to multiply the angular displacement expressed in radians by some radius r. The total angular stroke is 2γ radians, therefore, the total lineal stroke is

$$F = r \cdot 2\gamma \text{ and } r = \frac{F}{2\gamma} \text{ or } r = \frac{F}{2\gamma^\circ} \cdot \frac{180}{\pi} \text{ to have } \gamma \text{ in degrees.}$$

The velocity and the acceleration, as well as the displacement, are proportional to the stroke in the first power, therefore, all equations should be multiplied by r or its equivalent which permits to introduce the lineal stroke F. In cam equations, the working period is usually expressed in degrees what is reflected in the expression for the time T.

The introduction of α in that form can be made by multiplying the angular velocity equation by

$$\frac{360}{\alpha^\circ} \cdot \frac{2\psi^\circ}{360} = 1;$$

and the equation of acceleration by

$$\left(\frac{360}{\alpha^\circ} \cdot \frac{2\psi}{360}\right)^2 = 1$$

because the time enters in the second power.

Since angles α and ψ are presented as ratios, they can be in radians or in degrees.

The equations (3) and (4) express the velocity and the acceleration in radians. The transformation of them has the following aims:

(a) the equations should give the results in lineal measurements;

(b) all angles should be in degrees for convenient use;

(c) the equations should be written in dimensionless forms, developed for cams, so it would be possible to find dimensionless coefficients: C_V—for max. velocity and C_A—for maximum acceleration for an easy comparison of the performance of Geneva drives with performance of cams.

To satisfy aims (a) and (b), both equations should be multiplied by

$$r = \frac{F}{2\gamma} \cdot \frac{180}{\pi}$$

Besides, to satisfy the aim (c), the velocity equation (3) should be multiplied by

$$\frac{360}{\alpha} \cdot \frac{2\psi}{360}$$

and the equation (4) for the acceleration by

$$\left(\frac{360}{\alpha}\right)^2 \cdot \left(\frac{2\psi}{360}\right)^2$$

Since the speed N of the driver is already included in equations (3) and (4), it is possible to write the equation in the dimensionless forms—for velocity as $\frac{v}{F\frac{360}{\alpha}N}$ and acceleration as $\frac{a}{F\left(\frac{360}{\alpha}N\right)^2}$

Lineal velocity equation

$$v = w \cdot \frac{F \cdot 180}{2\gamma\pi} \cdot \frac{360 \cdot 2\psi}{\alpha \cdot 360} = w \cdot \left(F\frac{360}{\alpha^\circ}\right) \cdot \frac{\psi}{2\gamma\pi}$$

The angles ψ and γ can be expressed in terms of number of slots n.

$$v = w\left(F\frac{360}{\alpha^\circ}\right)\frac{1}{\pi} \cdot \frac{n-2}{4};$$

$$\frac{v}{F\frac{360}{\alpha^\circ}N} = \frac{n-2}{4} \cdot \frac{\frac{\sin\mu}{\tan\gamma} + \cos\mu - 1}{\frac{1}{2\tan^2\gamma} - \left(\frac{\sin\mu}{\tan\gamma} + \cos\mu - 1\right)} \qquad (3_1)$$

The chart 17 presents diagrams of the velocity for n = 4, 6, 8, 12, and 20.

The right side of the equation is dimensionless and its maximum value represents the coefficient of the maximum velocity—C_v. Cams have C_v with values between 1.6 and 2.0. Geneva drives have C_v in the same region, if $n \geqslant 6$ slots; C_v is higher for $n < 6$ slots.

To present diagrams of Geneva drives with different number of slots—the abcissa line shows fractions of the indexing period 2ψ, so it shows values between 0 and 1 for all indexing periods.

Lineal acceleration equation

$$a = \Omega \cdot \frac{F180}{2\gamma\pi} \cdot \left(\frac{360}{\alpha^\circ}\right)^2\left(\frac{2\psi^\circ}{360}\right)^2 = \Omega\left[F\left(\frac{360}{\alpha}\right)^2\right]\frac{\psi^2}{\gamma 360} \cdot \frac{1}{\pi};$$

$$a = \Omega\left[F\left(\frac{360}{\alpha^\circ}\right)\right]\frac{1}{\pi} \cdot \frac{(n-2)^2}{8n};$$

$$\frac{a}{F\left(\frac{360}{\alpha^\circ}N\right)^2} = \pi\,\frac{(n-2)^2}{8n} \cdot \frac{1}{\tan^2\gamma} \cdot \frac{\frac{\cos\mu}{\tan\gamma} - \sin\mu}{\left[\frac{1}{2\tan^2\gamma} - \frac{\sin\mu}{\tan\gamma} + \cos\mu - 1\right]^2} \qquad (4_1)$$

The right side of the equation is dimensionless and its maximum value represents the coefficient of the maximum acceleration—C_A.

The chart 18 represents diagrams for different n on the same unity base, considering $2\psi = 1$. The first observation is—all Geneva drives have a finite acceleration at the start and at the end of the indexing period, and it increases with the increase of number of slots n; it means a certain force is applied and removed instantaneously producing additional vibrational stresses. The coefficient C_A increases with a decrease of n and its position moves closer to the middle of the indexing period.

In the drives with three slots (n = 3) C_A is almost $16\frac{1}{2}$ and it changes to —16.5 in 0.15 of the indexing time; the violent change of the force will produce large vibrational stresses.

The drive with four slots (n = 4) has $C_A = +8.5$ and it changes to —8.5 in 0.25 of the indexing time.

Properly designed cams can have C_A between 4.9 and 5.5; in some cases $C_A = 6.3$, but in all cases, the acceleration starts and ends with zero value and changes gradually. Chart 18 has diagrams for Modified Trapezoid Acceleration with b = 0.15 and 0.25, but cams can be built with longer working periods α, than the working period of Geneva Drives; so the acceleration is decreased proportionally to the square of the increase of α.

It is evident that cams can be used at higher speed.

Geneva drives with 3 and 4 slots should be avoided and drives with $n \geqslant 6$ should be used at moderate speed because C_A at the start and at the end of the indexing period is $\geqslant 2.5$.

External Geneva drives have rather short indexing periods and that fact increases the acceleration, which is inversely proportional to the square of indexing time, while cams can have larger indexing periods for the same angular stroke of the wheel. This defect can be decreased considerably (4 times), if the driver has two rollers; unfortunately with larger n the locking of the wheel during the dwell period becomes less secure, if n is larger than 8.

For comparison, Modified Trapezoid Acceleration diagrams with b = 0.15 and b = 0.25 are also shown on the chart 18.

The table shows indexing periods of the Geneva Drive.

No. of slots n	3	4	5	6	8	9	10	12	15	20	30
$\alpha° = 2\psi$	60°	90°	108°	120°	135°	140°	144°	150°	156°	162°	168°

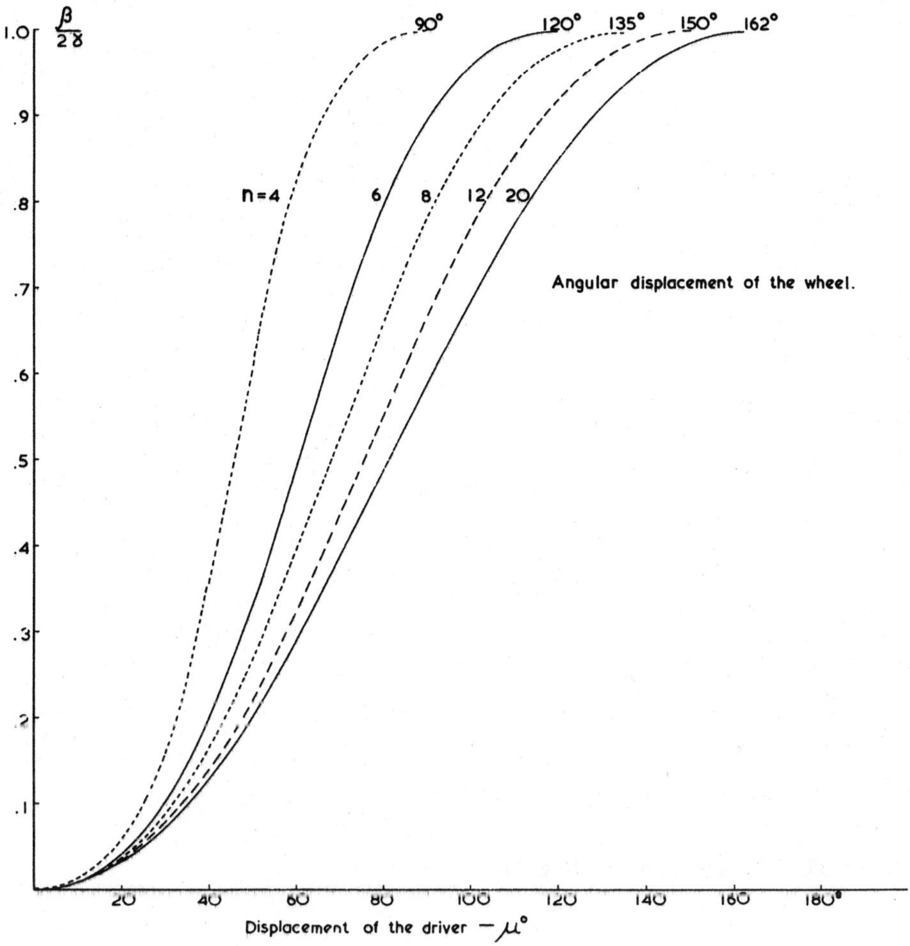

Chart 16. Geneva Drive—Regular

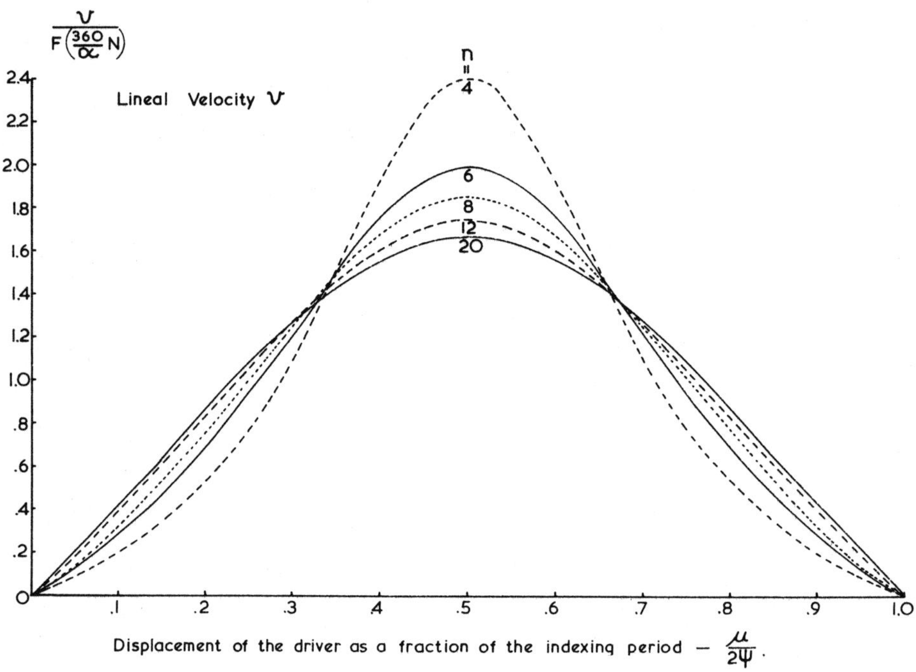

Chart 17. Geneva Drive—Regular

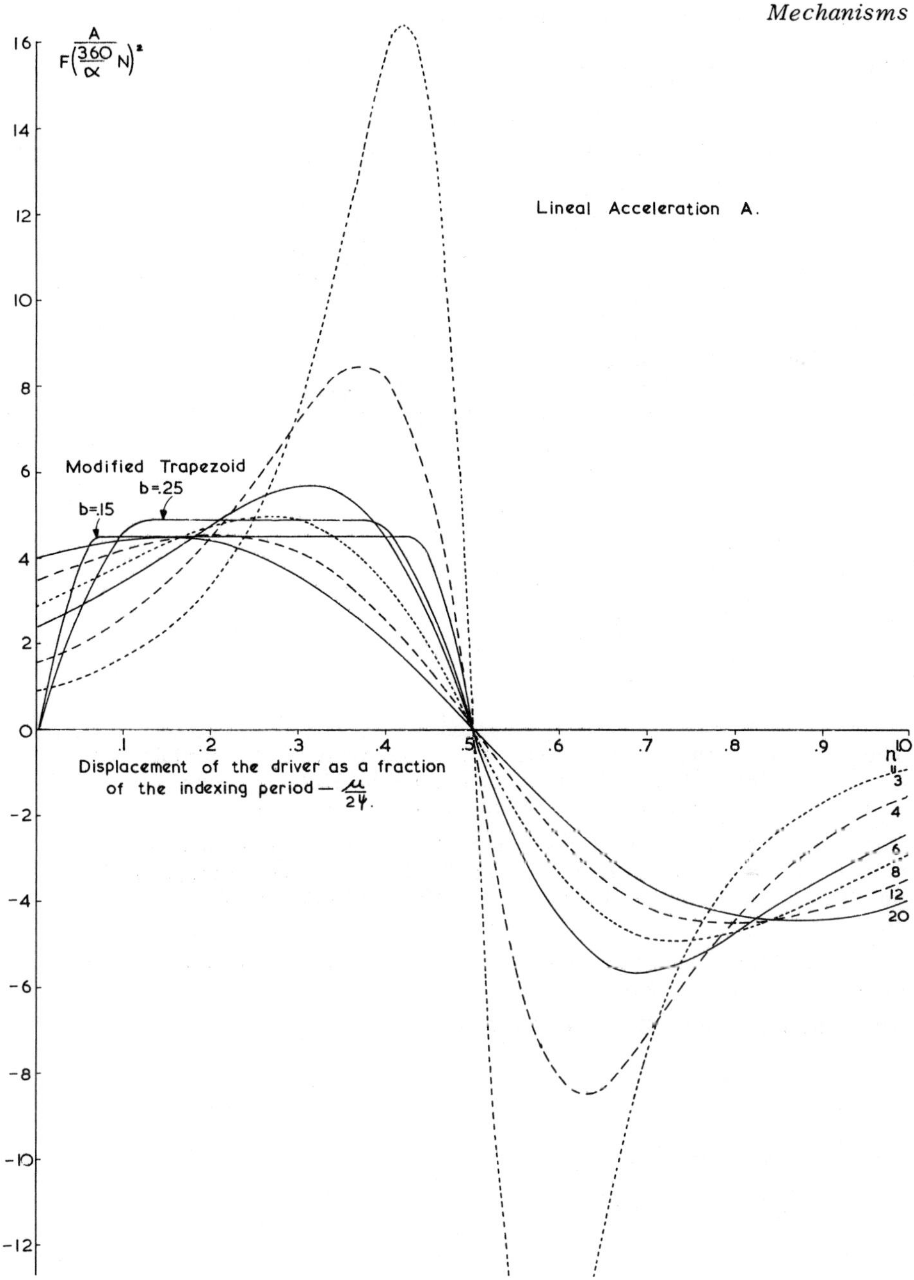

Chart 18. Geneva Drive—Regular

III b. Internal Geneva Drive

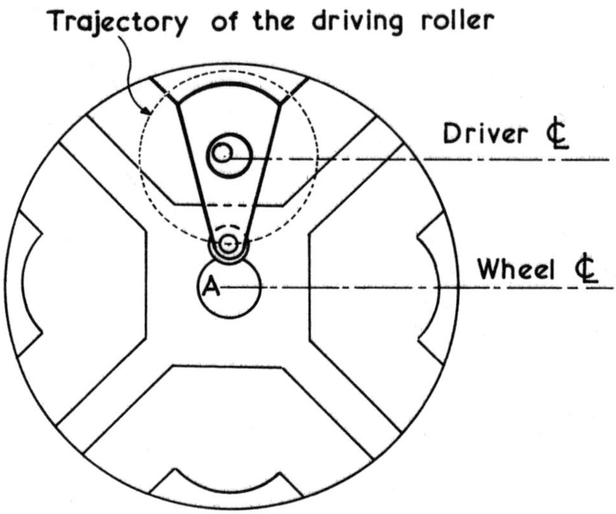

Fig. II - 12

The schematic drawing (Fig. II-12) shows the driver in the middle of a dwell. The drive shown on the drawing has four slots, but there may be 3 or more slots, as in the case of external Geneva drive.

For the analysis, the initial position of the driver is selected when the driver just enters the slot at the point E. The analyzed moment is when the driving roller moved from E to E_1, after the rotation of the driver on an angle μ

All points, dimensions and angles are marked by the same letters as in the case of external Geneva drives.

From the triangle OAE on the Fig. II-13

$$\gamma = \frac{180°}{n}$$

and $180° - \psi° = 90° - \gamma$

$$\psi = 90° + \gamma° = 90\frac{n+2}{n}$$

$$B = \frac{R}{\operatorname{Sin} \gamma} \qquad \text{(a)}$$

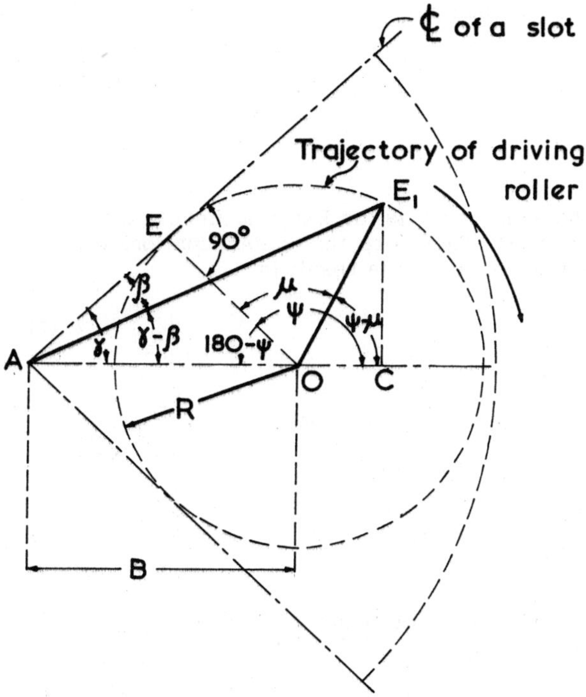

Fig. II - 13

From the triangle ACE_1

$$\tan(\gamma - \beta) = \frac{\overline{CE_1}}{\overline{AC}}$$

From the triangle OCE_1

$$\overline{CE_1} = R \operatorname{Sin}(\psi - \mu) \qquad \text{(b)}$$

$$\overline{OC} = R \operatorname{Cos}(\psi - \mu)$$

$$\overline{AC} = B + R \operatorname{Cos}(\psi - \mu) \qquad \text{(c)}$$

$$\tan(\gamma - \beta) = \frac{R \operatorname{Sin}(\psi - \mu)}{B + R \,.\, \operatorname{Cos}(\psi - \mu)} \qquad (1)$$

From the properties of the mechanism

$$\psi^\circ = 90^\circ + \gamma^\circ$$

$$\text{Sin } \psi = \text{Cos } \gamma \tag{d}$$

$$\text{and Cos } \psi = -\text{Sin } \gamma \tag{e}$$

All expressions are of the same form as for External Geneva drive, except for + and — signs in (c) and (e). Therefore, the transformations of the equation (1) are of the same nature, so the result is

$$\tan \beta = \frac{1 - \text{Cos}\,\mu}{\frac{1}{\tan \gamma} + \text{Sin } \mu} \tag{2}$$

The difference is only in + sign in the denominator.

The angular displacement of the wheel is

$$s = \text{arc tan } \frac{1 - \text{Cos } \mu}{\frac{1}{\tan \gamma} + \text{Sin } \mu} \text{ in radians} \tag{2_1}$$

The differentiation of the equation (2_1) is made in the same way as for External Geneva drive and

$$\beta = 2\pi \text{Nt} \tag{f}$$

The angular velocity as first derivative of the equation (2)

$$w = \pi N \frac{\frac{\text{Sin } \mu}{\tan \gamma} - \text{Cos } \mu + 1}{\frac{1}{2 \tan^2 \gamma} + \left(\frac{\text{Sin } \mu}{\tan \gamma} - \text{Cos } \mu + 1\right)} \tag{3}$$

The angular acceleration as the second derivative of the equation (2_1) or the first derivative of the equation (3)

$$\Omega = \frac{\pi^2 N^2}{\tan^2 \gamma} \cdot \frac{\frac{\text{Cos } \mu}{\tan \gamma} + \text{Sin } \mu}{\left[\frac{1}{2 \tan^2 \gamma} + \left(\frac{\text{Sin } \mu}{\tan \gamma} - \text{Cos } \mu + 1\right)\right]^2} \tag{4}$$

To transform the equations, expressed in radians, into equations for lineal velocity and acceleration—for comparison with cams, the same procedure is used as for External Geneva drive.

Lineal velocity

$$\frac{v}{F\dfrac{360}{\alpha^\circ}N} = \frac{n+2}{4} \cdot \frac{\dfrac{\text{Sin}\,\mu}{\tan\gamma} - \text{Cos}\,\mu + 1}{\dfrac{1}{2\tan^2\gamma} + \left(\dfrac{\text{Sin}\,\mu}{\tan\gamma} - \text{Cos}\,\mu + 1\right)} \qquad (3_1)$$

Lineal acceleration

$$\frac{a}{F\left(\dfrac{360}{\alpha^\circ}N\right)^2} = \pi \cdot \frac{(n+2)^2}{8n} \cdot \frac{1}{\tan\gamma} \cdot \frac{\dfrac{\text{Cos}\,\mu}{\tan\gamma} + \text{Sin}\,\mu}{\left[\dfrac{1}{2\tan^2\gamma} + \dfrac{\text{Sin}\,\mu}{\tan\gamma} - \text{Cos}\,\mu + 1\right]^2} \qquad (4_1)$$

The right sides of equations (3_1) and (4_1) are dimensionless and their maximum values give coefficients—C_V for maximum velocity and C_A for maximum acceleration. C_V are lower than in the case of cams; for different n it varies between 1.16 and 1.5. C_A are considerably higher and the maximum acceleration happens at the start and at the end of the indexing period.

Instantaneous application or removal of forces creats high vibrational stresses and they continue during the dwell period, so the next stroke may dangerously increase them.

Internal Geneva drives have long indexing periods, so the magnitude of dynamic force is considerably lower than in the case of External Geneva drive with the same number n of slots. In spite of that, the character of the acceleration limits their application only to low speed.

The chart 19 presents the fractional displacement of the wheel $\frac{\beta}{2\gamma}$ as a function of the angle of rotation μ of the driver.

The charts 20 and 21 present diagrams of the coefficients of the maximum lineal velocity and acceleration as a function of a fraction of the working period, which is accepted as unity.

The table shows working periods of the driver for different number of slots.

No. of slots n	3	4	5	6	8	9	10	12	15	20
Working period $\alpha^\circ = 2\psi$	300°	270°	252°	240°	225°	220°	216°	210°	204°	198°

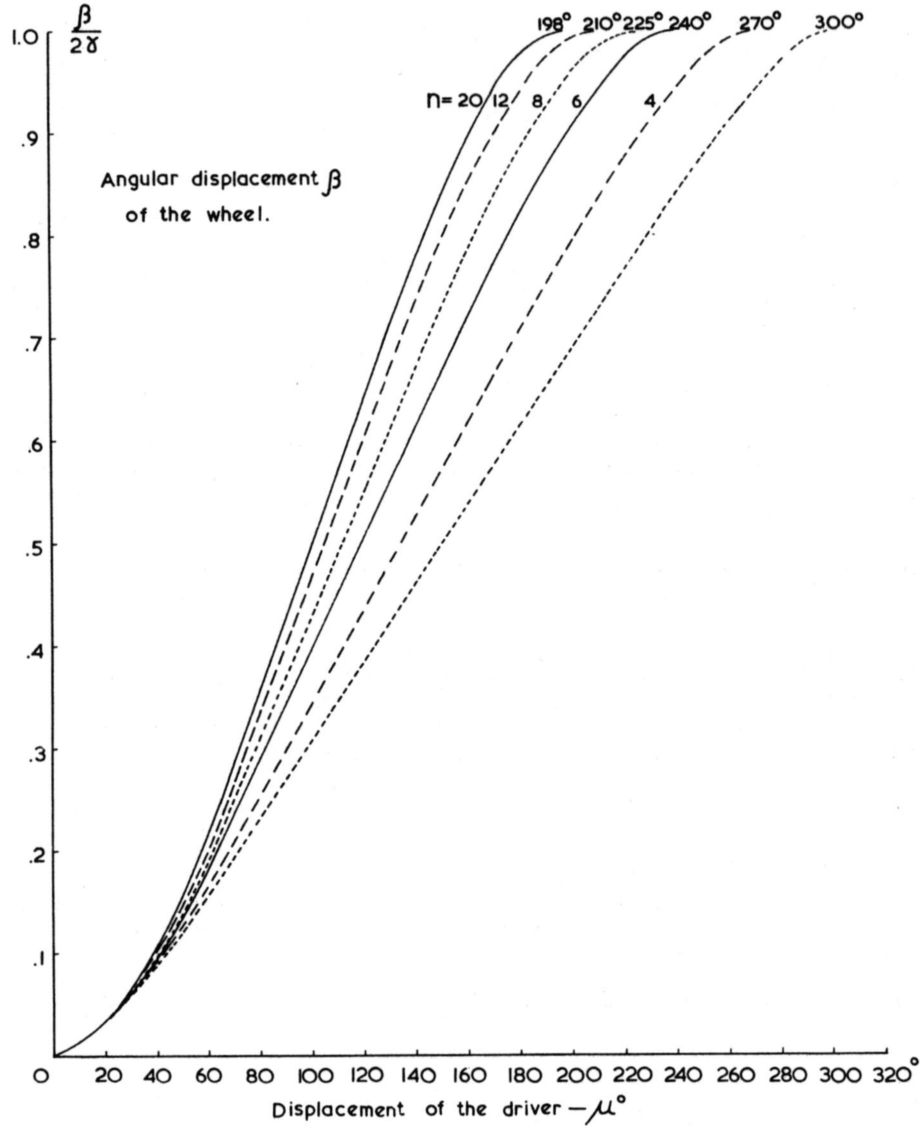

Chart 19. Geneva Drive—Internal

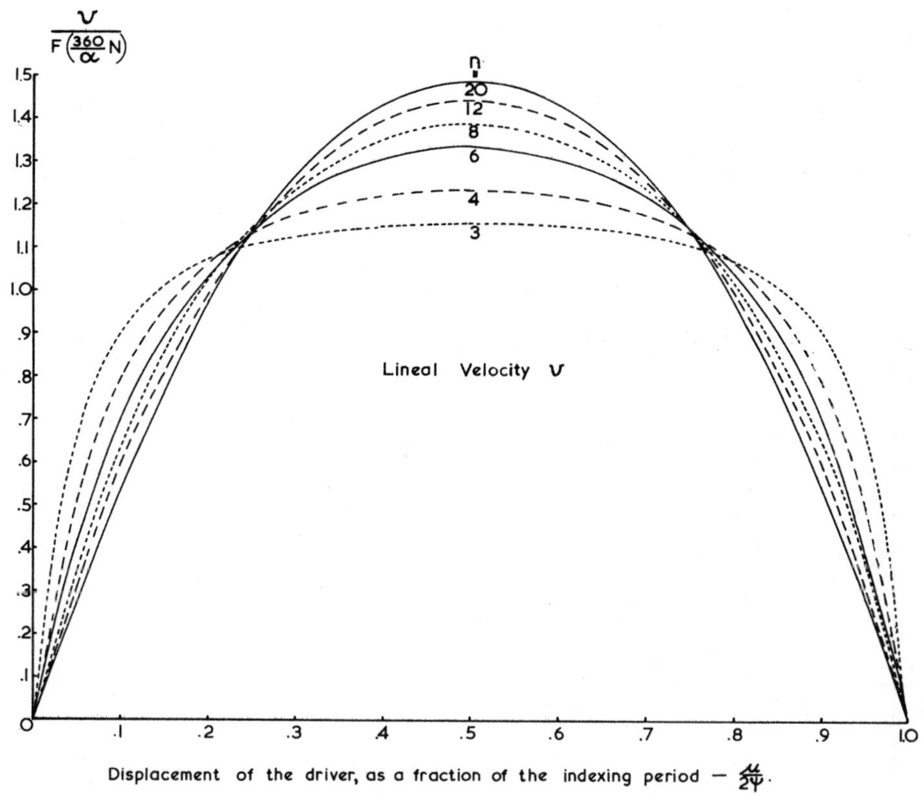

Chart 20. Geneva Drive—Internal

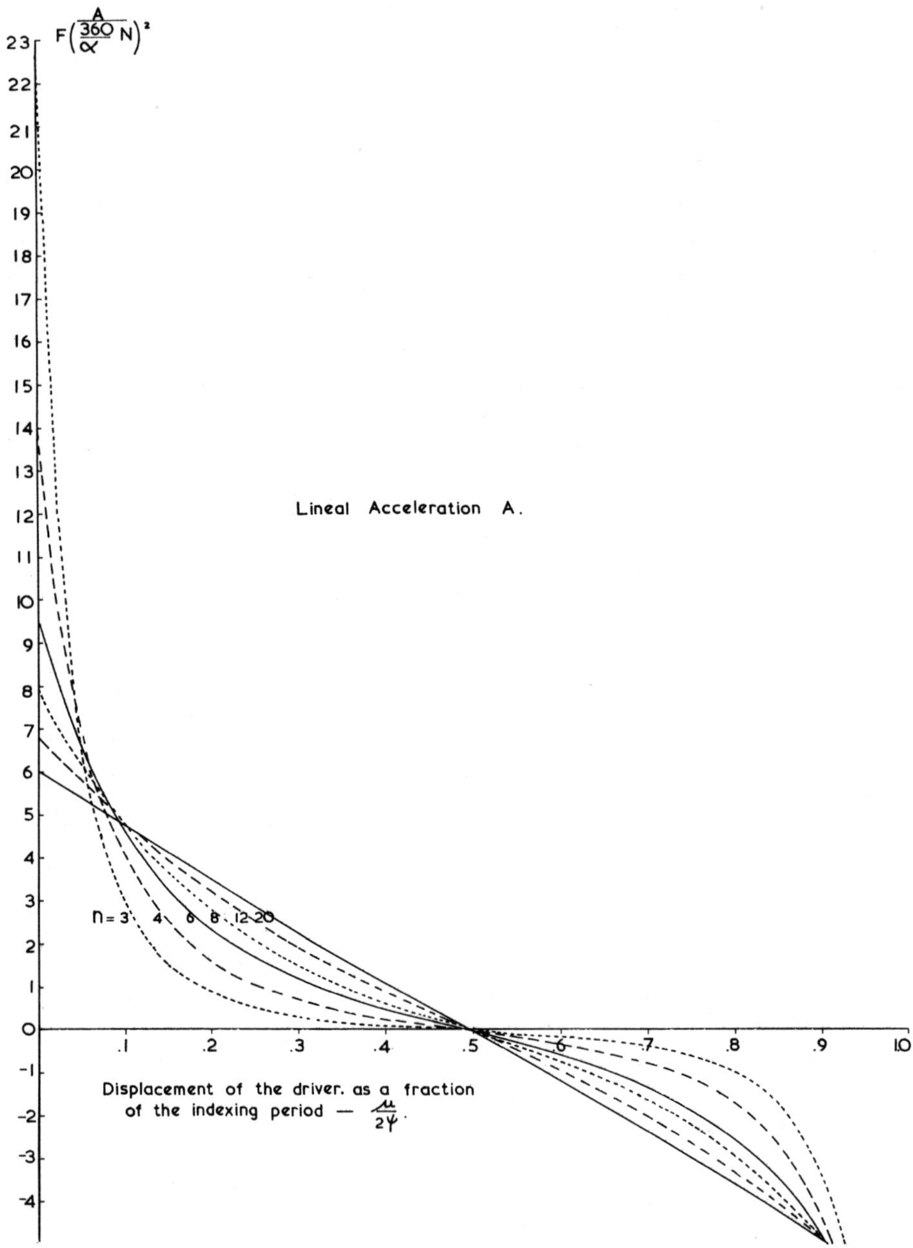

Chart 21. **Geneva Drive—Internal**

IIIc. Cam Driven Indexing Mechanisms

Indexing periods and acceleration diagrams can be selected by the designer. Driving cams have a barrel-like shape—with a groove or a rib to engage rollers attached to the driver wheel.

A grooved barrel cam permits a back-lash during the indexing period—when the direction of the action of dynamic force changes near the middle of the stroke. The cause of possible back-lash is in the necessity to have the roller diameter smaller than the width of a groove.

A barrel cam with a rib permits to use conjugate cam followers riding on both sides of the rib. If the rib has parallel sides, it is next to impossible to guarantee that both rollers are always in contact with the rib. Tapered ribs permit to preload rollers by a slight decrease of the distance between centers of the cam and of the wheel; in this case, it is possible to use cylindrical rollers with axes parallel to sides of the rib, so the precision of the roller position along its axis is immaterial. The use of tapered rollers against tapered ribs requires an extreme precision in positioning each roller along its axis and usually it is not practical.

Two basic designs of the driven wheel are used:

I. axes of rollers are parallel to the axis of the wheel (Fig. II-14).

II. rollers are located radially, so their axes are perpendicular to the axis of the wheel; this design was developed by the author and at the present time is known as Roller Gear Drive (Fig. II-15) or Ferguson Drive.

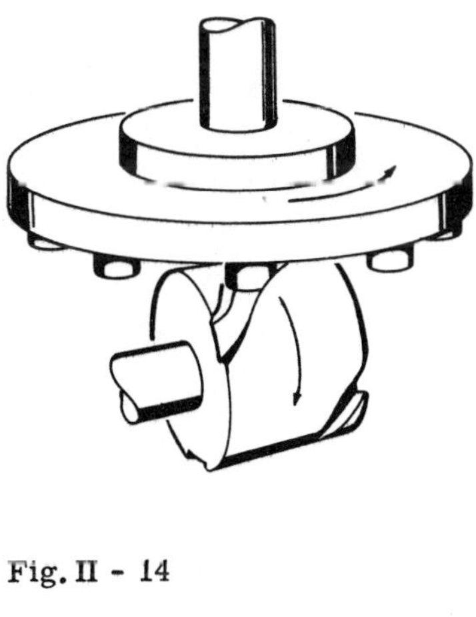

Fig. II - 14

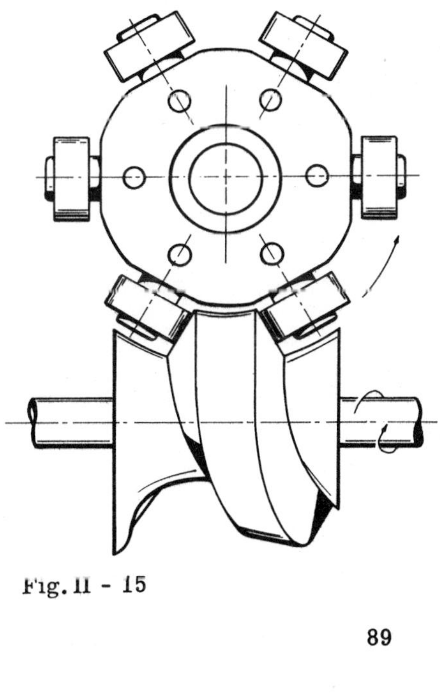

Fig. II - 15

The most practical design of the group I, Fig. II-14, has the movement of the wheel produced by one roller traveling in a groove with a changing helix angle; during the dwell two rollers are riding against the ends of the barrel cam; this design permits a secure locked position of the wheel during the dwell, if the angular location of rollers on the wheel is precise.

Roller Gear drive of the group II (Fig. II-15) has a tapered rib on a cam with a concave outline. Two rollers are in contact with two sides of the rib during the whole cycle—indexing and dwell—so there is no back-lash; the rollers are cylindrical, so their action is not affected by the location of rollers along their axes; standard antifriction bearings can be used as rollers; the preload on rollers is made by a slight change of the distance between the cam shaft and the shaft of the wheel. The minimum number of rollers on the wheel is six; but the cam may have more than one thread, so it is possible to have even only one dwell per revolution of the wheel; usually the number of stops (dwells) per revolution of the wheel is three or more.

CHAPTER III

CAMS AND "UNIVERSAL" METHOD OF CAM CALCULATIONS

Introduction

There are many different designs of cams; the majority of cams in automatic machines belong to the following types:

A. Cam followers movement is perpendicular to the axis of the cam shaft;

(1) plate cams with followers riding on periphery of them;
(2) plate cams with followers riding in a groove in the face of them;

A-1. Plate Cam

A-2. Face Cam

Courtesy of Ferguson Machine Co., St. Louis, Missouri.

B. Cam followers move in a plane parallel to the axis of the cam shaft;

(1) end cams;
(2) barrel cams with a groove or a rib.

B-1. End Cam

B-2. Barrel Cam with groove

B-2. Barrel Cam with rib

Courtesy of Ferguson Machine Co., St. Louis, Missouri.

C. Cams producing a continuous or an intermittent circular motion of the driven part with a number of cam followers:

(1) cylindrical barrel cams;
(2) enveloping or hour-glass cams.

There are modifications caused by requirements to have conjugate cam followers against a certain surface of a cam.

The purpose of automatic machines in manufacturing processes—to hold down the cost in spite of rising cost of production labor; this trend very often justifies investments into complicated machinery. Progressively, machines are designed to do many operations, requiring many different motions at a highest possible speed. In meeting these trends, cams offer definite advantages in comparison with many other mechanisms, formerly used widely in automatic machines.

Cams offer flexibility in the selection of working periods, dwells, and other items; the use of them permits a better timing of interfering movements in a machine.

They offer a possibility to control dynamic forces which are necessary to produce movements of machine parts; as a result, they create chances to balance these forces, so the speed limit is reached due to more than one critical movement. Also, cams have some advantages over linkages, as they can produce desired movements precisely.

Due to latest developments, cams of high quality can be designed with a minimum amount of time; there is a trend to introduce cams in connection with linkages and to utilize advantages of conjugate cam followers. As it was mentioned previously, the search for an "ideal" cam was started in 1930 and during the next 25 years, three groups of accelerations were developed and proved in practical application. All of them are based on trigonometric functions—sine and cosine. The other method is the development of acceleration diagrams based on polynomial equations; but it required the use of computers to develop equations for each set of requirements—a costly operation suitable only for cams used as a part of mass manufactured products, as automotive engines and different appliances.

Automatic machines are manufactured in small quantities and each machine usually employs a number of cams, so there was a need for a method which permits to design cams quickly and economically. By necessity, the search for "ideal" acceleration diagrams should be continued, because the requirements vary, so no single solution is the best for all of them.

Groups of Acceleration Diagrams

For convenience of designing different cams, it was necessary to develop each acceleration diagram in three forms, because there are three basic movements from which any movement, even a complicated one, can be constructed.

Type I — the stroke starts from a dwell and ends in a dwell.

Type II — the forward stroke is followed by the return stroke, which in turn is followed by the next forward stroke, so there are no dwells.

Type III — the forward (or initial) stroke is started from a dwell, but at the end of this stroke, it is followed by the return stroke without any dwell between them; the return stroke ends in a dwell.

The acceleration diagram for the Type III movement is a combination of diagrams for the Type I and the Type II movements.

The practice introduced some modifications. In some cases, there may be a uniform velocity period during one complete cycle of cam action. Usually, it is between the acceleration and the deceleration, but also it may be some time during one of these periods.

During the uniform velocity period, the acceleration is zero, because the acceleration represents the change of velocity. Therefore, so far as the acceleration diagram is concerned, the uniform velocity period acts in the same way as a dwell. The influence of the uniform velocity period is analized in the Chapter IV, Example X.

The introduction of the uniform velocity period increases the coefficient C_A, thus it increases dynamic forces, but it decreases the coefficient C_V, so a smaller cam can be designed. In one case of intermittent motion, the dwell was replaced by a period of uniform velocity (very low); then only the displacement table was modified by the addition of increments created by this velocity; this case is described in the Chapter IV, Example VII.

Any complicated motion can be divided into a group of elementary strokes of the three basic types.

Type I is used in most cases;

Type II is used very seldom, as in the case of reciprocal motion with a period of uniform velocity during one of the strokes.

Type III is used quite often, but less than Type I.

This book presents three groups of acceleration diagrams; each of them is developed for all three types of movements: Type I, Type II and Type III. These groups are based on the following acceleration diagrams:

> Modified Trapezoid—a combination of sine curve quadrants and uniform acceleration; the diagrams are shown with $b = 0.25$.
>
> > When $b = 0$, the diagram is of uniform acceleration and when $b = 0.5$, it is a sine curve.
>
> Modified Sine—a combination of sine quadrants occupying different parts of the working period; the diagrams are shown with $c = 0.25$.
>
> > When $c = 0$, the diagram is of cosine curve because first and fourth quadrants disappear and, when $c = 0.5$, it is a sine curve, as in the case of Modified Trapezoid.

Sine Curve, which is equivalent to other diagrams when b or $c = 0.5$. Each group of these diagrams has its field of usefulness.

Generally speaking—

Modified Trapezoid group is preferred when a lower acceleration is important or when a higher C_V is desirable, as in the case of included uniform velocity period; due to a comparatively fast change from + max. to − max. acceleration, it requires a cam follower system with a high frequency of vibration.

Modified Sine group is preferred for a high output torque, because its cam shaft torque is lower; also, it permits to make grooved cams—barrel or flat—smaller.

Sine—would be advantageous at high speed in case of unsufficient stiffness of the cam follower system; also, it has an advantage of lower residue vibrations.

All diagrams are shown on Fig. III-1 with equal periods of the acceleration and of the deceleration. For this condition, values of dimensionless coefficients C_A and C_V are shown; for diagrams for movements of the Type III, the coefficients of C_D of the maximum deceleration are shown because they are different than C_A.

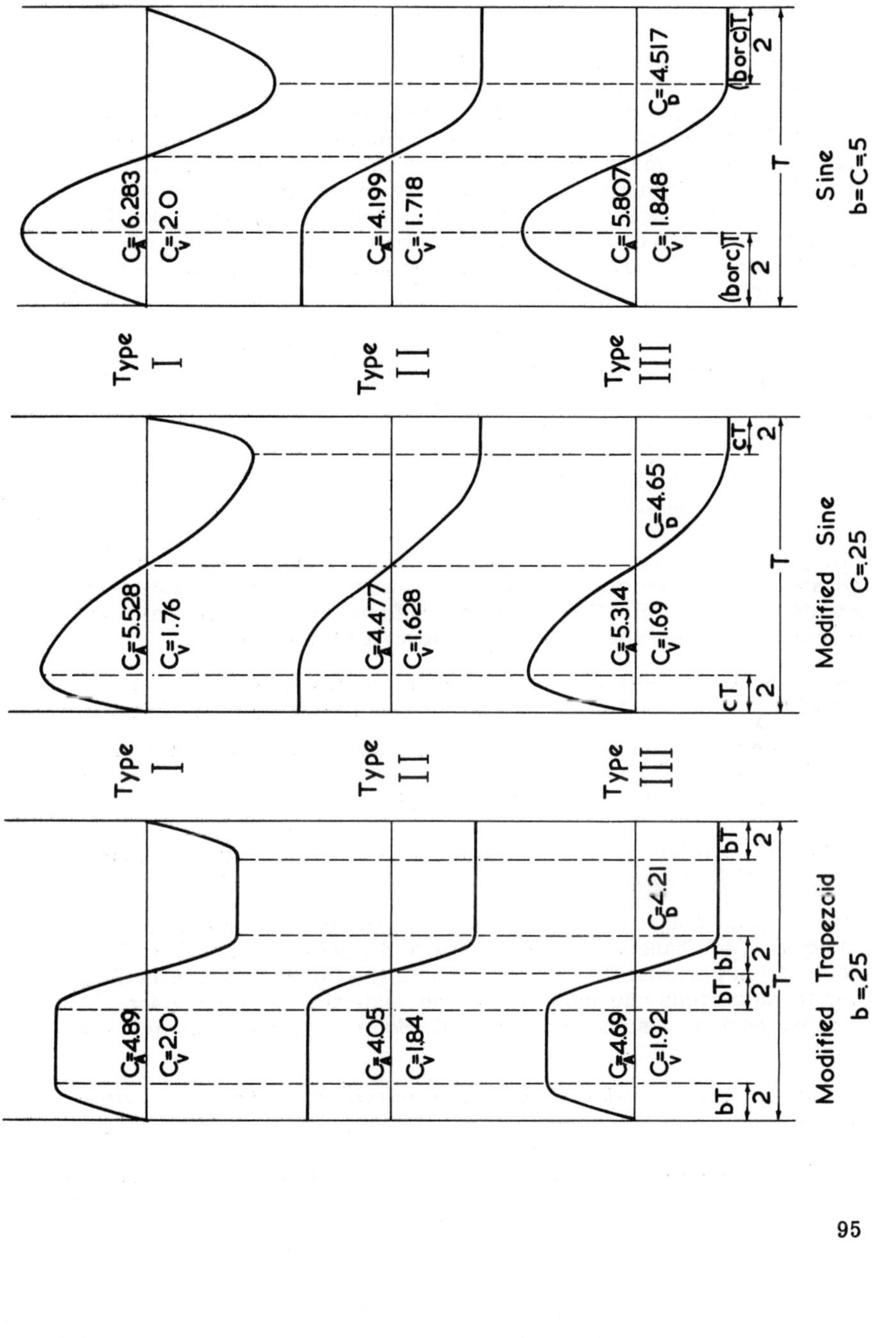
Type I
Type II
Type III
C_A=4.89
C_V=2.0
C_A=4.05
C_V=1.84
C_A=4.69
C_V=1.92
C_D=4.21
bT/2
T
Modified Trapezoid
b=.25
C_A=5.528
C_V=1.76
C_A=4.477
C_V=1.628
C_A=5.314
C_V=1.69
C_D=4.65
cT/2
Modified Sine
C=.25
C_A=6.283
C_V=2.0
C_A=4.199
C_V=1.718
C_A=5.807
C_V=1.848
C_D=4.517
(borc)T/2
Sine
b=C=.5

Fig. III - 1

Acceleration diagrams with the equal time for + and − acceleration and with definite ratios b and c were satisfactory for many years. With a progressive expansion of the field for automatic machines, more special cams were needed to satisfy peculiar and/or more precise requirements.

Any requirements can be satisfied by the designing special acceleration diagrams and preparing displacement tables for them. But machine designers in commercial concerns are pressed for time, when they have to design and build cams for manufacturing companies using a variety of automatic production machines. Most of the cams for special conditions are needed in small quantities, so a complete and original analysis of a problem cam and the preparation of displacement table would make the cost of cams prohibitive. So, the practice created the necessity to have flexible acceleration diagrams, which can be adapted to a quick solution of a variety of special cam problems.

The basis of the modification of acceleration diagrams was the introduction of a ratio, which can be easily varied by the designer.

Originally, the ratio was

$$P = \frac{\text{Time for the deceleration period}}{\text{Time for the acceleration period}}$$

in addition to the permanent ratios b and c, which determine a general configuration of an acceleration diagram. It was retained for cams with movements of Types I and II, but a different formulation became necessary for movements of the Type III; it will be described later on.

A change of time alloted for a stroke affects the magnitude of the acceleration and of the velocity. It does not affect the unity tables of displacement which represent the dimensionless ratio $\frac{s}{S}$, because the time of the stroke does not enter into equation for displacements.

Therefore, "unity tables" can be used for any values of working periods and for speed N RPS of the cam shaft.

When the original diagrams and tables were created, the ratio P was always equal 1, so changes of α and N affected equally the acceleration and the deceleration periods. The introduction of the factor $P \neq 1$ affects the time for these periods unequally and at the same time displacements during these periods also became unequal. So, the second conclusion was reached.

The displacement tables should be made separately for the acceleration and for the deceleration periods.

It means the displacement tables should be made not for $\frac{s}{S}$, which changes from 0 to 1, but separately for ratios $\frac{s}{S_A}$ and $\frac{s}{S_D}$;

S_A — the stroke during the acceleration period;
S_D — the stroke during the deceleration period;
S — the total stroke $= S_A + S_D$

The introduction of the ratio P affects the distribution of the time between + and — acceleration, but it does not change the shapes of diagrams, because the relations between elements of each part of a diagram—acceleration and deceleration—remain unchanged, due to the permanent factors b and c, which represent ratios.

So, the third conclusion is reached. The dimensionless displacement tables do not change and the tables, made for $P = 1$, can be used when $P \neq 1$.

The displacements S_A and S_D should be used for the multiplication of the dimensionless data from displacement tables. The actual displacement during the acceleration period s_A should vary between 0 and S_A, and s_D should vary between 0 and S_D.

The last conclusion is that—the separate dimensionless displacement tables for the acceleration and for the deceleration should be "unity tables," varying from 0 and 1, so the data can be multiplied by S_A or S_D.

Description of Displacement Tables for Movements of Type I and Type II

These movements have identical shapes for + and — acceleration when $P = 1$. It is evident that

$$S_A = S_D = \frac{S}{2}$$

The original tables were made for $\frac{s}{S}$ and $P = 1$, so at the end of + acceleration period $\frac{s}{S}$ is equal to 0.5. To make a separate unity table, it was necessary to double the figures in the original tables. To make a separate unity table for the deceleration period, it was necessary to subtract 0.5 from the figures in the original table and to double the result. When a table for the deceleration is multiplied by S_D, the results should be increased, adding S_A to every one of them.

All equations representing the acceleration, the velocity, and the displacement were changed by the introduction of the factor P, which changed the time for separate elements of the acceleration diagrams.

It was found that S_A, S_D, C_A, and C_D can be determined from simple expressions, showing the influence of the factor P. Also, it was found that the maximum velocity coefficient C_V is independent of the factor P.

The original displacement tables had 360 divisions for the whole stroke, so the separate tables have 180 divisions in each.

Description of Displacement Tables for Movements of Type III

The acceleration diagram is a combination of two shapes: one is the same as in the case of movements of the Type I, and for convenience, it is called the Shape I; the other is as in the case of movements of the Type II, so it is called the Shape II.

When the analysis was made, the Shape II represented the deceleration period and the Shape I represented the acceleration period, so

$$P = \frac{\text{Period occupied by Shape II}}{\text{Period occupied by Shape I}}$$

In the application of this diagram, either shape can represent the acceleration or the deceleration, but the value of P should be expressed as shown.

Again, the factor P changes the time allotted for both shapes, but the proportions of both curves remain the same as in diagrams for Type I or Type II movements. Therefore, dimensionless displacements remain the same, so the displacement tables for Types I and II movements can be used. Naturally, the presence of two different shapes affects the values of S_A and S_D.

To avoid any confusion, due to a more complicated character of the acceleration diagram, a different way of writing coefficients and displacements is offered.

- C_I — the coefficient of maximum acceleration attached to the Shape I and it may represent the maximum + or — acceleration.
- C_{II} — the coefficient of maximum acceleration attached to the Shape II and it may be a coefficient of maximum acceleration or of maximum deceleration.

The letters designating the displacements during two periods with different shapes of the acceleration diagrams (+ or —) should have two subscripts;

first—a subscript I or II, depending on the shape of the diagram which produces the displacement;

second—a subscript A or D, if the displacement is produced during the acceleration or during the deceleration.

The combination of subscripts indicates which table of displacement should be used. For instance, the subscript I-A indicates that the displacement S_{IA} should be used to multiply figures of the dimensionless displacement table marked I-A; the subscript I-D indicates that the displacement S_{ID} should be used with the data in the displacement table I-D. The same reasoning is applied to S_{IIA} and S_{IID}, so in these cases, the displacement tables for the diagram of the Type II are used.

In short, the distinction between acceleration and deceleration is used only in reference to the displacement tables; the coefficients of maximum acceleration C_I and C_{II} and the value of the factor P depend only on the shape of the acceleration diagram.

C_V — the coefficient of the maximum velocity is independent of the fact which shape is used for the acceleration or for the deceleration, but it is affected by the factor P.

The designer should be careful in calculating P, according to its definition for movements of the Type III.

It is advisable, at least in the beginning, to draw diagrams and to mark curves as on Fig. III-2.

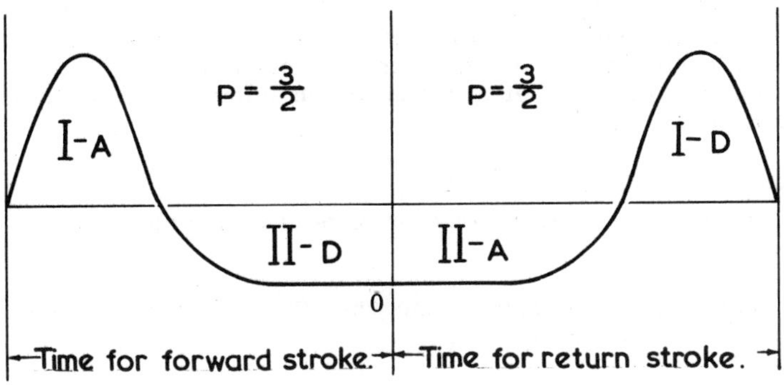

Fig. III - 2

The expressions for working periods α_I and α_{II} are as simple as in the case of movements of Types I and II and their subscripts indicate to what shape they belong.

Examples V and VI show how to balance accelerations of two strokes at the point 0 when working periods of two strokes are not equal.

Conclusion

The equations for working periods, coefficients and displacements are given on the next pages, also the short velocity tables and displacement tables. The information is given for three groups of acceleration diagrams and each group has information for movement of Types I, II, and III.

The described method was given the name:

"Universal" System of Cam Calculations

The name reflects the fact that the development of the method was done at Universal Match Corp., but also it meets requirements for universal application. It is a practical method, convenient for designers of automatic machines. It has

(1) Only 4 precalculated tables of displacements are required for movements of Types I, II and III of a certain group of diagrams.

(2) This method can be used without access to computers—only calculators are necessary to prepare displacement tables for the machine shop, though the tables can be programmed for computers.

(3) The designer can find quickly maximum values of + and − acceleration or velocity for different factors P.

(4) Displacements S_A and S_D (or S_I and S_{II}) are easily determined for the selected value of the factor P, so the designer can check positions of machine parts for some points of their strokes, where an interference between adjoining parts can be expected.

Process of Development of Equations for Acceleration, Velocity and Displacement

Sine curve represents the simplest case when P = 1, because one equation is used for the whole diagram.

$$\alpha = A \text{ Sin } 360°x = A \text{ Sin } 2\pi x \qquad (1)$$

When x varies from 0 to 1, Sin 360° varies from 0 to +1, then to 0 and to −1 and finally to 0. Since Sine has the maximum value ±1, A represents the maximum value of ± acceleration.

The variable x can be replaced by the ratio $\frac{t}{T}$, where t is the time which changes from 0 to T and T is the time of the stroke; therefore, this ratio varies in the same way as x. If the speed of the cam is N revolutions per second, the time of one revolution is $\frac{1}{N}$ sec. During the stroke, the cam rotates on $\alpha°$ or on $\alpha°/360$ part of one revolution, so

$$T = \frac{\alpha°}{360\ N} \text{ sec.}$$

To simplify the writing of equations, this expression will be introduced into final equations at the completion of the analysis.

To make the analysis more general and to demonstrate some complications, assume $P \neq 1$, so the time of the acceleration period is different from the time of the deceleration period. It means the time t varies between different limits during the acceleration and during the deceleration. The Fig. III-3 shows this fact clearly.

$$P = \frac{T_D}{T_A} \quad \text{and} \quad T_A + T_D = T.$$

From these equations, the values of T_A and T_D are found

$$T_A = \frac{1}{1 + P} \cdot T \quad \text{and} \quad T_D = \frac{P}{1 + P} \cdot T$$

During the acceleration period, the time t changes from 0 to $\frac{1}{1 + P} \cdot T$ and during the deceleration period, the time t changes from $\frac{1}{1 + P} \cdot T$ to T.

The variable x will have different forms for these two periods.

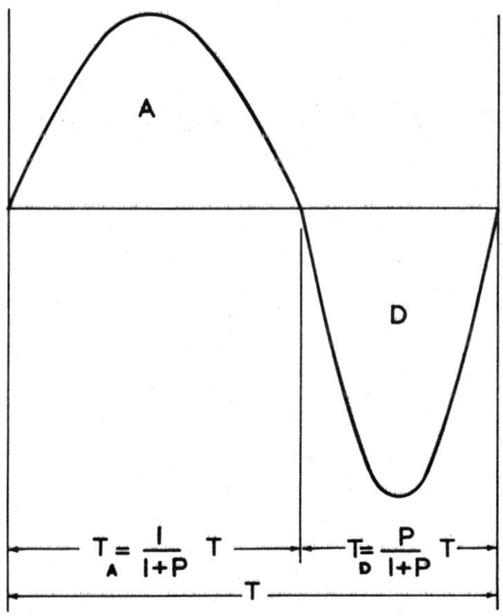

Fig. III - 3

During the acceleration period, x should change from 0 to ½ and during the deceleration, it should change from ½ to 1, because trigonometric functions are used for the acceleration diagram,—so when x changes from 0 to ½ the value of Sin 360°x changes from 0 to +1 and back to 0. The variations of time are different. Here is one of the possible methods for finding the forms for x. Since x and $\frac{t}{T}$ vary proportionally, a first power can be used

$$Ux = \frac{t}{T} - Z \quad \text{and} \quad x = \frac{\frac{t}{T} - Z}{U}$$

where U and Z are unknown.

The values of x and $\frac{t}{T}$ are known for the beginning and for the end of each period, so this equation with two unknown items can be easily solved and x can be expressed, as a function of $\frac{t}{T}$; the use of these expressions becomes clear in the following application of them.

Acceleration equations

Acceleration period: x changes from 0 to ½; while $\frac{t}{T}$ changes from 0 to $\frac{1}{1 + P}$.

At the beginning of the period $U \cdot 0 = 0 - Z; \quad Z = 0$

At the end of it,

$$U \cdot \frac{1}{2} = \frac{1}{1 + P} - 0; \quad U = \frac{2}{1 + P}; \quad x = \frac{\frac{t}{T} - 0}{\frac{2}{1 + P}} = \frac{1 + P}{2} \cdot \frac{t}{T}$$

$$a_1 = A \operatorname{Sin} 2\pi \cdot \frac{1 + P}{2} \cdot \frac{t}{T} \qquad (1_1)$$

Deceleration period: x changes from ½ to 1; while $\frac{t}{T}$ changes from $\frac{1}{1 + P}$ to 1;

At the beginning:

$$U \cdot \frac{1}{2} = \frac{1}{1+P} - Z; \qquad Z = \frac{1}{1+P} - \frac{U}{2}$$

At the end:

$$U \cdot 1 = 1 - Z = 1 - \frac{1}{1+P} + \frac{U}{2} = \frac{P}{1+P} + \frac{U}{2}; \qquad \frac{U}{2} = \frac{P}{1+P};$$

$$U = \frac{2P}{1+P};$$

$$Z = \frac{1}{1+P} - \frac{P}{1+P} = \frac{1-P}{1+P};$$

$$x = \frac{\left(\frac{t}{T} - \frac{1-P}{1+P}\right)\left(1+P\right)}{2P} = \frac{1+P}{2P} \cdot \frac{t}{T} - \frac{1-P}{2P}$$

It is not correct to retain A for both periods, it should be replaced by the maximum deceleration value D for the deceleration period.

$$a_2 = D \operatorname{Sin} 2\pi \left(\frac{1+P}{2P} \cdot \frac{t}{T} - \frac{1-P}{2P}\right) \qquad (1_2)$$

Velocity equations.

$$v_1 = \int_0^t a_1 \cdot dt = A\int_0^t \operatorname{Sin} 2\pi \cdot \frac{1+P}{2} \cdot \frac{t}{T} \cdot dt =$$

$$= A\left[-\frac{\operatorname{Cos} 2\pi \frac{1+P}{2} \cdot \frac{t}{T}}{2\pi \frac{1+P}{2T}}\right]_0^t = \frac{AT}{2\pi} \cdot \frac{2}{1+P}\left[-\operatorname{Cos} 2\pi \frac{1+P}{2} \cdot \frac{t}{T}\right]_0^t;$$

when $t = 0; -\operatorname{Cos} 2\pi \frac{1+P}{2} \cdot \frac{t}{T} = -\operatorname{Cos} 0 = -1$

$$v_1 = \frac{AT}{2\pi} \cdot \frac{2}{1+P} \cdot \left(1 - \operatorname{Cos} 2\pi \frac{1+P}{2} \cdot \frac{t}{T}\right) \qquad (2_1)$$

when $t = \frac{1}{1+P}$ T, at the end of the acceleration period,

$$-\operatorname{Cos} 2\pi \frac{1+P}{2} \cdot \frac{t}{T} = -\operatorname{Cos} \pi = +1;$$

and

$$v_I = \frac{AT}{2\pi} \cdot \frac{4}{1+P} \tag{2'_1}$$

This velocity should be added to the varying velocity during the deceleration period; also v_I is the maximum velocity $= v_{max}$.

$$v_2 = v_I + \int_{\frac{T}{1+P}}^{t} a_2 \, . \, dt = v_1 + D \int_{\frac{T}{1+P}}^{t} \operatorname{Sin} 2\pi\left(\frac{1+P}{2P} \cdot \frac{t}{T} - \frac{1-P}{2P}\right) . \, dt =$$

$$= v_I + D\left[-\frac{\operatorname{Cos} 2\pi\left(\frac{1+P}{2P} \cdot \frac{t}{T} - \frac{1-P}{2P}\right)}{\frac{2\pi}{T} \cdot \frac{1+P}{2P}}\right]_{\frac{T}{1+P}}^{t} =$$

$$= v_I + \frac{DT}{2\pi} \cdot \frac{2P}{1+P}\left[-\operatorname{Cos} 2\pi\left(\frac{1+P}{2P} \cdot \frac{t}{T} - \frac{1-P}{2P}\right)\right]_{\frac{T}{1+P}}^{t};$$

when $t = \frac{T}{1+P}$; $\quad -\operatorname{Cos} 2\pi\left(\frac{1+P}{2P} \cdot \frac{t}{T} - \frac{1-P}{2P}\right) = -\operatorname{Cos} \pi = +1$

$$v_2 = v_I + \frac{DT}{2\pi} \cdot \frac{2P}{1+P}\left[-1 - \operatorname{Cos} 2\pi\left(\frac{1+P}{2P} \cdot \frac{t}{T} - \frac{1-P}{2P}\right)\right] =$$

$$= \frac{T}{2\pi} \cdot \frac{2}{1+P}\left[A \, . \, 2 - DP\left(1 + \operatorname{Cos} 2\pi\left(\frac{1+P}{2P} \cdot \frac{t}{T} - \frac{1-P}{2P}\right)\right)\right] \tag{2_2}$$

v_{II} at the end of the stroke should be equal zero.

When t = T; $\quad \text{Cos } 2\pi\left(\frac{1+P}{2P} \cdot \frac{t}{T} - \frac{1-P}{2P}\right) = \text{Cos } 2\pi = +1;$

$$v_{II} = \frac{T}{2\pi} \cdot \frac{2}{1+P} (A \cdot 2 - DP \cdot 2) = 0.$$

It can happen when A = DP or $D = \frac{A}{P}$ $\quad (2_2)$

Then the velocity equation will be

$$v_2 = \frac{AT}{2\pi} \cdot \frac{2}{1+P}\left[1 - \text{Cos } 2\pi\left(\frac{1+P}{2P} \cdot \frac{t}{T} - \frac{1-P}{2P}\right)\right] \qquad (2_2)$$

Displacement equations.

$$s_1 = \int_0^t v_1 \cdot dt = \frac{AT}{2\pi} \cdot \frac{2}{1+P} \int_0^t \left(1 - \text{Cos } 2\pi \frac{1+P}{2} \cdot \frac{t}{T}\right) \cdot dt =$$

$$= \frac{AT^2}{2\pi} \cdot \frac{2}{1+P}\left[\frac{t}{T} - \frac{2}{2\pi(1+P)} \text{Sin } 2\pi \frac{1+P}{2} \cdot \frac{t}{T}\right]_0^t$$

When t = 0, the whole expression in parantheses is equal zero.

$$s_1 = \frac{AT^2}{2\pi} \cdot \frac{2}{1+P}\left(\frac{t}{T} - \frac{2}{2\pi(1+P)} \text{Sin } 2\pi \frac{1+P}{2} \cdot \frac{t}{T}\right) \qquad (3_1)$$

At the end of the acceleration period $t = \frac{1}{1+P} \cdot T$

and

$$\text{Sin } 2\pi \frac{1+P}{2} \cdot \frac{t}{T} = \text{Sin } \pi = 0$$

So, $s_I = \frac{AT^2}{2\pi} \cdot \frac{2}{(1+P)^2}$ $\quad (3_1)$

This displacement should be added to the displacements during the deceleration period.

$$s_2 = s_I + \int_{\frac{T}{1+P}}^{t} v_2 \, . \, dt = s_I + \frac{AT}{2\pi} \cdot \frac{2}{1+P} \int_{\frac{T}{1+P}}^{t}$$

$$\left(1 - \text{Cos}\, 2\pi\left(\frac{1+P}{2P} \cdot \frac{t}{T} - \frac{1-P}{2P}\right)\right) . \, dt =$$

$$= s_I + \frac{AT^2}{2\pi} \cdot \frac{2}{1+P}\left[\frac{t}{T} - \frac{2P}{2\pi(1+P)} \text{Sin}\, 2\pi\left(\frac{1+P}{2P} \cdot \frac{t}{T} - \frac{1-P}{2P}\right)\right]_{\frac{T}{1+P}}^{t};$$

when $t = \frac{T}{1+P}$; $\quad \text{Sin}\, 2\pi\left(\frac{1+P}{2P} \cdot \frac{t}{T} - \frac{1-P}{2P}\right) = \text{Sin}\, \pi = 0$ and

$\frac{t}{T} = \frac{1}{1+P}$;

$$s_2 = s_I + \frac{AT^2}{2\pi} \cdot \frac{2}{1+P}\left(\frac{t}{T} - \frac{2P}{2\pi(1+P)} \text{Sin}\, 2\pi\left(\frac{1+P}{2P} \cdot \frac{t}{T} - \frac{1-P}{2P}\right) - \frac{1}{1+P}\right) =$$

$$= \frac{AT^2}{2\pi} \cdot \frac{2}{1+P}\left[\frac{t}{T} - \frac{2P}{2\pi(1+P)} \text{Sin}\, 2\pi\left(\frac{1+P}{2P} \cdot \frac{t}{T} - \frac{1-P}{2P}\right)\right] \qquad (3_2)$$

At the end of the deceleration period

$$t = T; \quad \text{Sin}\, 2\pi\left(\frac{1+P}{2P} \cdot \frac{t}{T} - \frac{1-P}{2P}\right) = \text{Sin}\, 2\pi = 0;$$

also, the displacement s_{II} will give the full stroke S.

$$s_{II} = S = \frac{AT^2}{2\pi} \cdot \frac{2}{1+P} \qquad (3_2)$$

$$A = \frac{S}{T^2} \cdot 2\pi \frac{1+P}{2} \quad \text{and} \quad D = \frac{A}{P} = \frac{S}{T^2} 2\pi \frac{1+P}{2P} \qquad (4)$$

The coefficients of maximum acceleration are affected by the factor P.

$$C_A = 2\pi \frac{1+P}{2} \quad \text{and} \quad C_D = 2\pi \frac{1+P}{2P} \qquad (5)$$

The expression

$$\frac{S}{T^2} = S\left(\frac{360}{\alpha^\circ} N\right)^2$$

A and D should be introduced into all equations, which should be written in dimensionless form as

$$\frac{a}{\frac{S}{T^2}} \text{ or } \frac{a}{S\left(\frac{360}{\alpha^\circ} N\right)^2}; \quad \frac{v}{\frac{S}{T}} \text{ or } \frac{v}{S \cdot \frac{360}{\alpha^\circ} N} \quad \text{and} \quad \frac{s}{S}$$

To find the coefficient of the maximum velocity, introduce A from the equation (4) into the equation (2_1) for $v_I = v_{max}$

$$v_{max} = 2\pi \frac{1 + P}{2} \cdot \frac{S}{T^2} \frac{T}{2\pi} \cdot \frac{4}{1 + P} = 2 \frac{S}{T}$$

therefore

$$C_v = 2 \qquad (6)$$

and it is not affected by the factor P.

The next pages give tables of displacements for three groups of acceleration diagrams. Each group has tables for movements of the Types I and II, and they are used also for the movements of the Type III.

Under each of the nine diagrams are equations describing different items as functions of the factor P; they are used extensively in the preparation of displacement tables for the manufacturing of cams.

Short tables for the velocity are supplied, as fractions of the maximum velocity tied up to the same divisions as in unity displacement tables.

DIAGRAM I

Modified Trapezoid; b = 0.25

The stroke is performed between dwells.

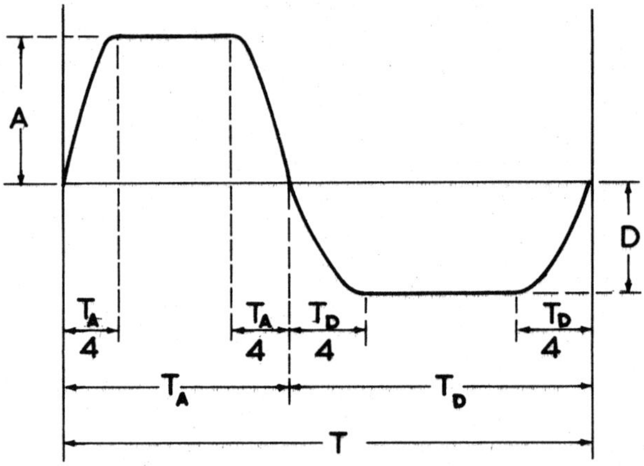

Fig. III - 4

Equations:

$$T_A = \frac{1}{1+P} \cdot T; \quad S_A = \frac{1}{1+P} \cdot S; \qquad \alpha_A = \frac{1}{1+P} \alpha.$$

$$T_D = \frac{P}{1+P} T; \quad S_D = \frac{P}{1+P} \cdot S; \qquad \alpha_D = \frac{P}{1+P} \alpha.$$

$$A_{max} = C_A \cdot \frac{S}{T^2}; \; C_A = 4.8881 \frac{1+P}{2};$$

$$D_{max} = C_D \cdot \frac{S}{T^2}; \; C_D = 4.8881 \frac{1+P}{2P}; \; \frac{C_A}{C_D} = P$$

$$V_{max} = C_V \cdot \frac{S}{T}; \quad C_V = 2;$$

$$T = \frac{\alpha^\circ}{360N};$$

Steps in calculating displacement table for a specific cam

(1) Select P and check C_A and C_D;

(2) Calculate S_A and S_D;

(3) Multiply Table I-A by S_A;

(4) Multiply Table I-D by S_D and add S_A to every item.

Velocity V_{max}

During Acceleration			During Acceleration		
0	0.0000	180	96	0.5407	84
6	0.0043	174	102	0.5815	78
12	0.0168	168	108	0.6222	72
18	0.0372	162	114	0.6629	66
24	0.0644	156	120	0.7037	60
30	0.0973	150	126	0.7444	54
36	0.1344	144	132	0.7851	48
42	0.1742	138	138	0.8258	42
48	0.2149	132	144	0.8656	36
54	0.2556	126	150	0.9027	30
60	0.2963	120	156	0.9356	24
66	0.3371	114	162	0.9628	18
72	0.3778	108	168	0.9832	12
78	0.4185	102	174	0.9957	6
84	0.4593	96	180	1.0000	0
90	0.5000	90			
		During Deceleration			During Deceleration

MODIFIED TRAPEZOID, DIAGRAM I ; b = .25

I-A ACCELERATION PERIOD

0	000 000								
1	000 000	37	020 447	73	125 416	109	328 132	145	628 572
2	000 004	38	022 049	74	129 727	110	335 158	146	638 297
3	000 012	39	023 724	75	134 113	111	342 260	147	648 091
4	000 028	40	025 473	76	138 575	112	349 438	148	657 954
5	000 055	41	027 296	77	143 112	113	356 691	149	667 885
6	000 095	42	029 194	78	147 725	114	364 018	150	677 883
7	000 150	43	031 166	79	152 413	115	371 422	151	687 946
8	000 224	44	033 214	80	157 177	116	378 901	152	698 073
9	000 318	45	035 337	81	162 016	117	386 456	153	708 262
10	000 436	46	037 536	82	166 930	118	394 086	154	718 512
11	000 580	47	039 810	83	171 920	119	401 791	155	728 823
12	000 752	48	042 160	84	176 985	120	409 572	156	739 191
13	000 954	49	044 585	85	182 126	121	417 429	157	749 615
14	001 189	50	047 085	86	187 342	122	425 361	158	760 093
15	001 461	51	049 661	87	192 633	123	433 367	159	770 624
16	001 770	52	052 313	88	198 000	124	441 450	160	781 204
17	002 119	53	055 040	89	203 443	125	449 608	161	791 833
18	002 509	54	057 842	90	208 961	126	457 842	162	802 509
19	002 944	55	060 720	91	214 554	127	466 151	163	813 230
20	003 426	56	063 672	92	220 223	128	474 535	164	823 993
21	003 956	57	066 701	93	225 966	129	482 995	165	834 794
22	004 537	58	069 805	94	231 786	130	491 530	166	845 634
23	005 170	59	072 984	95	237 681	131	500 140	167	856 510
24	005 858	60	076 239	96	243 652	132	508 826	168	867 420
25	006 601	61	079 569	97	249 698	133	517 588	169	878 358
26	007 402	62	082 975	98	255 819	134	526 425	170	889 325
27	008 263	63	086 456	99	262 016	135	535 337	171	900 319
28	009 184	64	090 013	100	268 287	136	544 325	172	911 335
29	010 168	65	093 645	101	274 635	137	553 389	173	922 372
30	011 216	66	097 352	102	281 058	138	562 527	174	933 428
31	012 330	67	101 135	103	287 557	139	571 740	175	944 501
32	013 510	68	104 993	104	294 131	140	581 028	176	955 584
33	014 758	69	108 927	105	300 780	141	590 390	177	966 679
34	016 074	70	112 936	106	307 505	142	599 827	178	977 781
35	017 461	71	117 021	107	314 305	143	609 336	179	988 889
36	018 918	72	121 181	108	321 181	144	618 918	180	1.000 000

MODIFIED TRAPEZOID, DIAGRAM I; b = .25

I-D DECELERATION PERIOD

0	000 000								
1	011 111	37	390 664	73	685 695	109	882 979	145	982 539
2	022 219	38	400 173	74	692 495	110	887 064	146	983 926
3	033 321	39	409 610	75	699 220	111	891 073	147	985 242
4	044 416	40	418 972	76	705 869	112	895 007	148	986 490
5	055 499	41	428 260	77	712 443	113	898 865	149	987 670
6	066 572	42	437 473	78	718 942	114	902 648	150	988 784
7	077 628	43	446 611	79	725 365	115	906 355	151	989 832
8	088 665	44	455 675	80	731 713	116	909 987	152	990 816
9	099 681	45	464 663	81	737 984	117	913 544	153	991 737
10	110 675	46	473 575	82	744 181	118	917 025	154	992 598
11	121 642	47	482 412	83	750 302	119	920 431	155	993 399
12	132 580	48	491 174	84	756 348	120	923 761	156	994 142
13	143 490	49	499 860	85	762 319	121	927 016	157	994 830
14	154 366	50	508 470	86	768 214	122	930 195	158	995 463
15	165 206	51	517 005	87	774 034	123	933 299	159	996 044
16	176 007	52	525 465	88	779 777	124	936 328	160	996 574
17	186 770	53	533 849	89	785 446	125	939 280	161	997 056
18	197 491	54	542 158	90	791 039	126	942 158	162	997 491
19	208 167	55	550 392	91	796 557	127	944 960	163	997 881
20	218 796	56	558 550	92	802 000	128	947 687	164	998 230
21	229 376	57	566 633	93	807 367	129	950 339	165	998 539
22	239 907	58	574 639	94	812 658	130	952 915	166	998 811
23	250 385	59	582 571	95	817 874	131	955 415	167	999 046
24	260 809	60	590 428	96	823 015	132	957 840	168	999 248
25	271 177	61	598 209	97	828 080	133	960 190	169	999 420
26	281 488	62	605 914	98	833 070	134	962 464	170	999 564
27	291 738	63	613 544	99	837 984	135	964 663	171	999 682
28	301 927	64	621 099	100	842 823	136	966 786	172	999 776
29	312 054	65	628 578	101	847 587	137	968 834	173	999 850
30	322 117	66	635 982	102	852 275	138	970 806	174	999 905
31	332 115	67	643 309	103	856 888	139	972 704	175	999 945
32	342 046	68	650 562	104	861 425	140	974 527	176	999 972
33	351 909	69	657 740	105	865 887	141	976 276	177	999 988
34	361 703	70	664 842	106	870 273	142	977 951	178	999 996
35	371 428	71	671 868	107	874 584	143	979 553	179	1.000 000
36	381 082	72	678 819	108	878 819	144	981 082	180	1.000 000

DIAGRAM II

Modified Trapezoid; b = 0.25

The movement is reversed without a dwell.

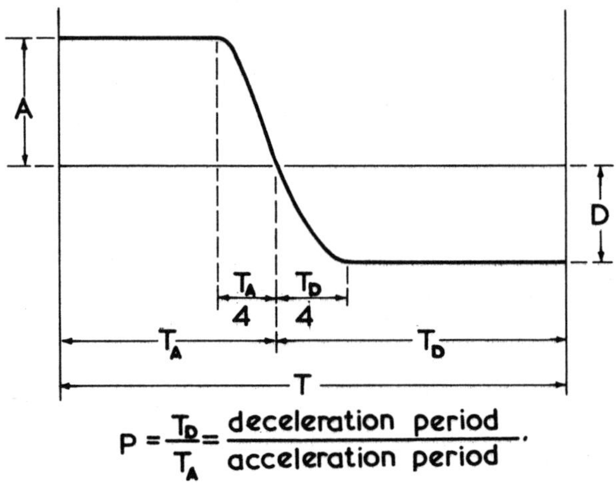

Fig. III - 5

Equations:

$$T_A = \frac{1}{1+P} \cdot T; \quad S_A = \frac{1}{1+P} S; \quad \alpha_A = \frac{1}{1+P} \alpha;$$

$$T_D = \frac{P}{1+P} \cdot T; \quad S_D = \frac{P}{1+P} S; \quad \alpha_D = \frac{P}{1+P} \alpha;$$

$$A = C_A \cdot \frac{S}{T^2}; \quad C_A = 4.0479 \frac{1+P}{2};$$

$$D = C_D \cdot \frac{S}{T^2}; \quad C_D = 4.0479 \frac{1+P}{2P}; \quad \frac{C_A}{C_D} = P$$

$$V_{max} = C_V \cdot \frac{S}{T}; \quad C_V = 1.8401$$

Steps in calculating displacement table for a specific cam

(1) Select P and check C_A and C_D;

(2) Calculate S_A and S_D;

(3) Multiply Table II-A by S_A;

(4) Multiply Table II-D by S_D and add S_A to every item.

Velocity
V_{max}

During Acceleration			During Acceleration		
0	0.0000	180	96	0.5867	84
6	0.0367	174	102	0.6233	78
12	0.0733	168	108	0.6600	72
18	0.1100	162	114	0.6967	66
24	0.1467	156	120	0.7333	60
30	0.1833	150			
36	0.2200	144	126	0.7700	54
42	0.2567	138	132	0.8067	48
48	0.2933	132	138	0.8433	42
54	0.3300	126	144	0.8791	36
60	0.3667	120	150	0.9125	30
66	0.4033	114	156	0.9421	24
72	0.4400	108	162	0.9666	18
78	0.4767	102	168	0.9849	12
84	0.5133	96	174	0.9962	6
90	0.5500	90	180	1.0000	0
		During Deceleration			During Deceleration

MODIFIED TRAPEZOID, DIAGRAM II; b = .25

II-A ACCELERATION PERIOD

0	000 000								
1	000 031	37	042 759	73	166 446	109	371 091	145	656 663
2	000 125	38	045 102	74	171 037	110	377 931	146	665 738
3	000 281	39	047 507	75	175 691	111	384 834	147	674 870
4	000 500	40	049 974	76	180 408	112	391 799	148	684 060
5	000 781	41	052 504	77	185 187	113	398 827	149	693 305
6	001 124	42	055 097	78	190 028	114	405 917	150	702 606
7	001 530	43	057 752	79	194 932	115	413 070	151	711 961
8	001 999	44	060 469	80	199 898	116	420 285	152	721 368
9	002 530	45	063 249	81	204 927	117	427 562	153	730 828
10	003 123	46	066 091	82	210 017	118	434 902	154	740 338
11	003 779	47	068 996	83	215 171	119	442 305	155	749 897
12	004 498	48	071 963	84	220 387	120	449 770	156	759 504
13	005 279	49	074 993	85	225 666	121	457 297	157	769 158
14	006 122	50	078 085	86	231 007	122	464 887	158	778 857
15	007 028	51	081 240	87	236 410	123	472 539	159	788 599
16	007 996	52	084 457	88	241 876	124	480 254	160	798 382
17	009 027	53	087 736	89	247 404	125	488 031	161	808 205
18	010 120	54	091 078	90	252 995	126	495 871	162	818 068
19	011 275	55	094 483	91	258 649	127	503 773	163	827 967
20	012 493	56	097 950	92	264 365	128	511 738	164	837 901
21	013 774	57	101 479	93	270 143	129	519 765	165	847 868
22	015 117	58	105 071	94	275 984	130	527 854	166	857 867
23	016 523	59	108 726	95	281 887	131	536 006	167	867 895
24	017 991	60	112 443	96	287 852	132	544 221	168	877 949
25	019 521	61	116 222	97	293 880	133	552 498	169	888 030
26	021 114	62	120 064	98	299 971	134	560 838	170	898 134
27	022 770	63	123 968	99	306 124	135	569 240	171	908 259
28	024 488	64	127 935	100	312 340	136	577 704	172	918 403
29	026 268	65	131 964	101	318 618	137	586 231	173	928 565
30	028 111	66	136 056	102	324 958	138	594 820	174	938 742
31	030 016	67	140 209	103	331 361	139	603 471	175	948 932
32	031 984	68	144 426	104	337 827	140	612 184	176	959 132
33	034 014	69	148 705	105	344 355	141	620 959	177	969 341
34	036 106	70	153 047	106	350 945	142	629 795	178	979 557
35	038 261	71	157 451	107	357 598	143	638 691	179	989 778
36	040 479	72	161 917	108	364 313	144	647 648	180	1.000 000

MODIFIED TRAPEZOID, DIAGRAM II; b = .25

II-D DECELERATION PERIOD

0	000 000								
1	010 222	37	361 309	73	642 402	109	842 549	145	961 739
2	020 443	38	370 205	74	649 055	110	846 953	146	963 894
3	030 659	39	379 041	75	655 645	111	851 295	147	965 986
4	040 868	40	387 816	76	662 173	112	855 574	148	968 016
5	051 068	41	396 529	77	668 639	113	859 791	149	969 984
6	061 258	42	405 180	78	675 042	114	863 944	150	971 889
7	071 435	43	413 769	79	681 382	115	868 036	151	973 732
8	081 597	44	422 296	80	687 660	116	872 065	152	975 512
9	091 741	45	430 760	81	693 876	117	876 032	153	977 230
10	101 866	46	439 162	82	700 029	118	879 936	154	978 886
11	111 970	47	447 502	83	706 120	119	883 778	155	980 479
12	122 051	48	455 779	84	712 148	120	887 557	156	982 009
13	132 105	49	463 994	85	718 113	121	891 274	157	983 477
14	142 133	50	472 146	86	724 016	122	894 929	158	984 883
15	152 132	51	480 235	87	729 857	123	898 521	159	986 226
16	162 099	52	488 262	88	735 635	124	902 050	160	987 507
17	172 033	53	496 227	89	741 351	125	905 517	161	988 725
18	181 932	54	504 129	90	747 005	126	908 922	162	989 880
19	191 795	55	511 969	91	752 596	127	912 264	163	990 973
20	201 618	56	519 746	92	758 124	128	915 543	164	992 004
21	211 401	57	527 461	93	763 590	129	918 760	165	992 972
22	221 143	58	535 113	94	768 993	130	921 915	166	993 878
23	230 842	59	542 703	95	774 334	131	925 007	167	994 721
24	240 496	60	550 230	96	779 613	132	928 037	168	995 502
25	250 103	61	557 695	97	784 829	133	931 004	169	996 221
26	259 662	62	565 098	98	789 983	134	933 909	170	996 877
27	269 172	63	572 438	99	795 073	135	936 751	171	997 470
28	278 632	64	579 715	100	800 102	136	939 531	172	998 001
29	288 039	65	586 930	101	805 068	137	942 248	173	998 470
30	297 394	66	594 083	102	809 972	138	944 903	174	998 876
31	306 695	67	601 173	103	814 813	139	947 496	175	999 219
32	315 940	68	608 201	104	819 592	140	950 026	176	999 500
33	325 130	69	615 166	105	824 309	141	952 493	177	999 719
34	334 262	70	622 069	106	828 963	142	954 898	178	999 875
35	343 337	71	628 909	107	833 554	143	957 241	179	999 969
36	352 352	72	635 687	108	838 083	144	959 521	180	1.000 000

DIAGRAM III

Modified Trapezoid; b = 1/4

The stroke starts from a dwell but is reversed without a dwell.

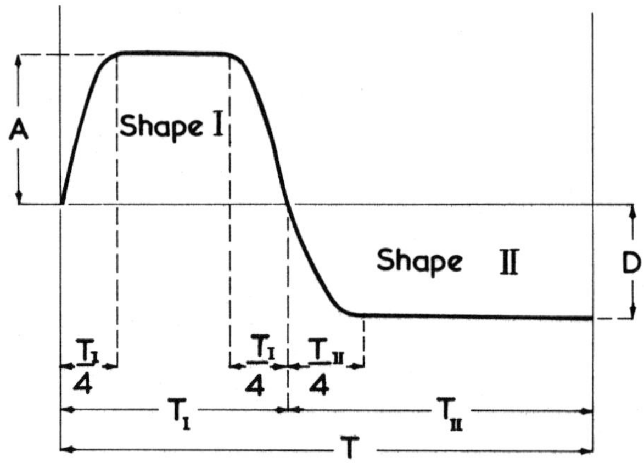

Fig. III - 6

Equations:

$$T_I = \frac{1}{1+P} \cdot T; \qquad S_{IA \text{ or } ID} = \frac{1}{1 + 1.0869\,P}\,S; \qquad \alpha_I = \frac{1}{1+P}\,\alpha;$$

$$T_{II} = \frac{P}{1+P} \cdot T; \qquad S_{IID \text{ or } II.A} = \frac{1.0869\,P}{1 + 1.0869\,P}\,S;$$

$$\alpha_{II} = \frac{P}{1+P}\,\alpha;$$

$$D_I \text{ or } A_I = C_I \frac{S}{T^2}; \qquad C_I = 4.8881\left(\frac{1+P}{2}\right)\left(\frac{1+P}{1 + 1.0869\,P}\right)$$

$$A_{II} \text{ or } D_{II} = C_{II} \frac{S}{T^2}; \qquad C_{II} = 4.3907\left(\frac{1+P}{2P}\right)\left(\frac{1+P}{1 + 1.0869\,P}\right);$$

$$\frac{C_I}{C_{II}} = 1.111\,P$$

$$V_{max} = C_V \cdot \frac{S}{T}; \qquad C_V = 2\,\frac{1 + P}{1 + 1.0869\ P};$$

$$T = \frac{\alpha^\circ}{360N};$$

Tables for the displacement are the same as for Diagrams I and II, according to the subscripts I-A; I-D; II-A; II-D.

The diagram is a combination of two shapes: one—from Diagram I and the other—from Diagram II.

Either one of them may represent the acceleration period.

In this case the ratio P has different meaning than in cases of symmetrical Diagrams I and II.

$$P = \frac{\text{Period occupied by the Shape II (acceleration or deceleration)}}{\text{Period occupied by the Shape I}} =$$

$$= \frac{T_{II}}{T_I}$$

Displacements S, time T, acceleration A and deceleration D have subscripts different than used for symmetrical diagrams I and II.

The subscripts I and II indicate that the item belongs to a certain shape (I or II).

Displacements S have additional subscripts—A and D—designating that shapes I or II are used for acceleration or for deceleration periods; these subscripts also indicate to which unity table the displacement should be applied.

The diagram, as shown above, has S_{IA} and S_{IID}; so S_{IA} should be used with the unity displacement Table I-A and S_{IID}—with the unity displacement Table II-D.

The tables for velocities are used in a similar way.

DIAGRAM I

Modified Sine; $c = 0.25$

The stroke is performed between dwells.

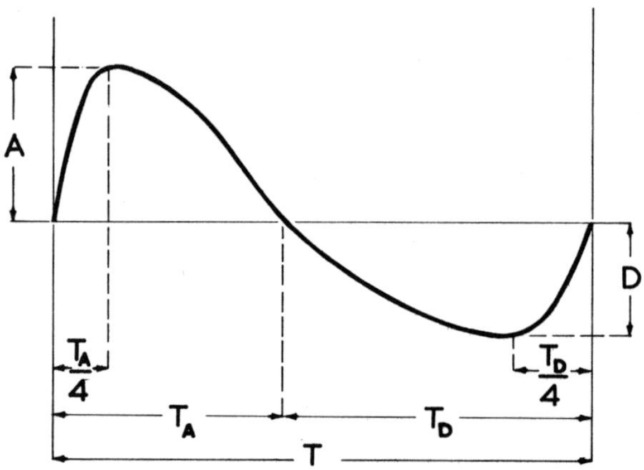

$$P = \frac{T_D}{T_A} = \frac{\text{deceleration period}}{\text{acceleration period}}.$$

Fig. III - 7

Equations:

$$T_A = \frac{1}{1 + P} T; \quad S_A = \frac{1}{1 + P} S; \quad \alpha_A = \frac{1}{1 + P} \alpha;$$

$$T_D = \frac{P}{1 + P} T; \quad S_D = \frac{P}{1 + P} S; \quad \alpha_D = \frac{P}{1 + P} \alpha;$$

$$A = C_A \frac{S}{T^2}; \quad C_A = 5.5280 \frac{1 + P}{2}$$

$$\frac{C_A}{C_D} = P.$$

$$D = C_D \cdot \frac{S}{T^2}; \quad C_D = 5.5280 \frac{1 + P}{2P}$$

$$V_{max} = C_V \cdot \frac{S}{T}; \quad C_V = 1.7596$$

$$T = \frac{\alpha^\circ}{360N};$$

Steps in calculating displacement table for a specific cam

(1) Select P and check C_A and C_D;

(2) Calculate S_A and S_D.

(3) Multiply Table I-A by S_A.

(4) Multiply Table I-D by S_D and add S_A to every item.

Velocity V_{max}

During Acceleration			During Acceleration		
0	0.0000	180	96	0.6694	84
6	0.0055	174	102	0.7117	78
12	0.0216	168	108	0.7518	72
18	0.0477	162	114	0.7895	66
24	0.0827	156	120	0.8245	60
30	0.1250	150			
36	0.1727	144	126	0.8568	54
42	0.2239	138	132	0.8860	48
48	0.2762	132	138	0.9122	42
54	0.3284	126	144	0.9352	36
60	0.3802	120	150	0.9548	30
66	0.4314	114	156	0.9709	24
72	0.4818	108	162	0.9836	18
78	0.5310	102	168	0.9927	12
84	0.5788	96	174	0.9982	6
90	0.6250	90	180	1.0000	0
		During Deceleration			During Deceleration

MODIFIED SINE, DIAGRAM I; c = .25

I-A ACCELERATION PERIOD

0	000 000								
1	000 000	37	023 123	73	141 538	109	363 157	145	664 889
2	000 004	38	024 935	74	146 368	110	370 601	146	674 082
3	000 014	39	026 829	75	151 279	111	378 107	147	683 308
4	000 032	40	028 807	76	156 271	112	385 674	148	692 565
5	000 062	41	030 869	77	161 342	113	393 302	149	701 854
6	000 107	42	033 015	78	166 493	114	400 990	150	711 172
7	000 170	43	035 246	79	171 723	115	408 738	151	720 520
8	000 253	44	037 562	80	177 032	116	416 544	152	729 896
9	000 360	45	039 963	81	182 419	117	424 408	153	739 299
10	000 493	46	042 449	82	187 884	118	432 330	154	748 729
11	000 656	47	045 021	83	193 426	119	440 306	155	758 184
12	000 851	48	047 678	84	199 045	120	448 339	156	767 664
13	001 079	49	050 420	85	204 742	121	456 426	157	777 167
14	001 346	50	053 248	86	210 514	122	464 568	158	786 693
15	001 653	51	056 161	87	216 362	123	472 763	159	796 240
16	002 001	52	059 159	88	222 286	124	481 010	160	805 808
17	002 396	53	062 242	89	228 285	125	489 310	161	815 396
18	002 839	54	065 410	90	234 357	126	497 660	162	825 002
19	003 330	55	068 662	91	240 503	127	506 059	163	834 626
20	003 875	56	071 999	92	246 723	128	514 509	164	844 267
21	004 475	57	075 421	93	253 016	129	523 007	165	853 924
22	005 131	58	078 928	94	259 381	130	531 553	166	863 595
23	005 847	59	082 519	95	265 818	131	540 146	167	873 280
24	006 624	60	086 194	96	272 327	132	548 785	168	882 978
25	007 465	61	089 953	97	278 907	133	557 469	169	892 688
26	008 371	62	093 796	98	285 555	134	566 197	170	902 409
27	009 344	63	097 722	99	292 274	135	574 969	171	912 140
28	010 386	64	101 732	100	299 062	136	583 783	172	921 880
29	011 499	65	105 825	101	305 916	137	592 640	173	931 627
30	012 684	66	110 001	102	312 842	138	601 538	174	941 381
31	013 943	67	114 260	103	319 834	139	610 474	175	951 142
32	015 278	68	118 602	104	326 890	140	619 451	176	960 908
33	016 689	69	123 026	105	334 014	141	628 465	177	970 678
34	018 178	70	127 532	106	341 203	142	637 517	178	980 450
35	019 746	71	132 119	107	348 457	143	646 606	179	990 224
36	021 394	72	136 788	108	355 775	144	655 730	180	1.000 000

MODIFIED SINE, DIAGRAM I ; c = .25

I-D

DECELERATION PERIOD

0	000 000								
1	009 776	37	353 394	73	651 543	109	867 881	145	980 254
2	019 550	38	362 483	74	658 797	110	872 468	146	981 822
3	029 322	39	371 535	75	665 986	111	876 974	147	983 311
4	039 092	40	380 549	76	673 110	112	881 398	148	984 722
5	048 858	41	389 526	77	680 166	113	885 740	149	986 057
6	058 619	42	398 462	78	687 158	114	889 999	150	987 316
7	068 373	43	407 360	79	694 084	115	894 175	151	988 501
8	078 120	44	416 217	80	700 938	116	898 268	152	989 614
9	087 860	45	425 031	81	707 726	117	902 278	153	990 656
10	097 591	46	433 803	82	714 445	118	906 204	154	991 629
11	107 312	47	442 531	83	721 093	119	910 047	155	992 535
12	117 022	48	451 215	84	727 673	120	913 806	156	993 376
13	126 720	49	459 854	85	734 182	121	917 481	157	994 153
14	136 405	50	468 447	86	740 619	122	921 072	158	994 869
15	146 076	51	476 993	87	746 984	123	924 579	159	995 525
16	155 733	52	485 491	88	753 277	124	928 001	160	996 125
17	165 374	53	493 941	89	759 497	125	931 338	161	996 670
18	174 998	54	502 340	90	765 643	126	934 590	162	997 161
19	184 604	55	510 690	91	771 715	127	937 758	163	997 604
20	194 192	56	518 990	92	777 714	128	940 841	164	997 999
21	203 760	57	527 237	93	783 638	129	943 830	165	998 347
22	213 307	58	535 432	94	789 486	130	946 752	166	998 654
23	222 833	59	543 574	95	795 258	131	949 580	167	998 921
24	232 336	60	551 661	96	800 955	132	952 322	168	999 149
25	241 816	61	559 694	97	806 574	133	954 979	169	999 344
26	251 271	62	567 670	98	812 116	134	957 551	170	999 507
27	260 701	63	575 592	99	817 581	135	960 037	171	999 640
28	270 104	64	583 456	100	822 968	136	962 438	172	999 747
29	279 480	65	591 262	101	828 277	137	964 754	173	999 830
30	288 828	66	599 010	102	833 507	138	966 985	174	999 893
31	298 146	67	606 698	103	838 658	139	969 131	175	999 938
32	307 435	68	614 326	104	843 729	140	971 193	176	999 968
33	316 692	69	621 893	105	848 721	141	973 171	177	999 986
34	325 918	70	629 399	106	853 632	142	975 065	178	999 996
35	335 111	71	636 843	107	858 462	143	976 877	179	1.000 000
36	344 270	72	644 225	108	863 212	144	978 606	180	1.000 000

DIAGRAM II

Modified Sine; $c = 0.25$

The movement is reversed without a dwell.

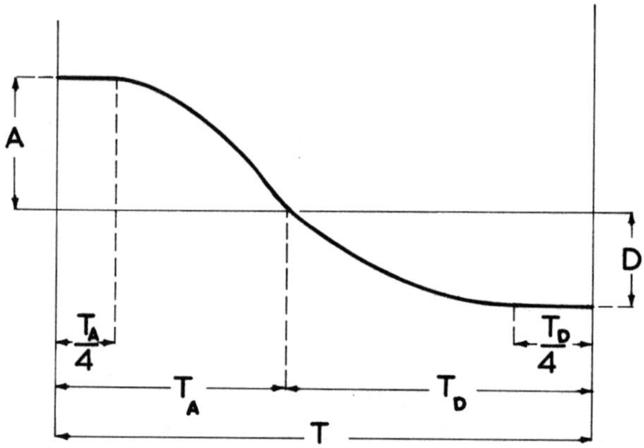

$$P = \frac{T_D}{T_A} = \frac{\text{deceleration period}}{\text{acceleration period}}.$$

Fig. III - 8

Equations:

$$T_A = \frac{1}{1+P}\,T; \qquad S_A = \frac{1}{1+P}\,S; \qquad \alpha_A = \frac{1}{1+P}\,\alpha;$$

$$T_D = \frac{P}{1+P}\,T; \qquad S_D = \frac{P}{1+P}\,S; \qquad \alpha_D = \frac{P}{1+P}\,\alpha;$$

$$A = C_A \cdot \frac{S}{T^2}; \qquad C_A = 4.4770\,\frac{1+P}{2}$$

$$\frac{C_A}{C_D} = P$$

$$D = C_D \cdot \frac{S}{T^2}; \qquad C_D = 4.4770\,\frac{1+P}{2P}$$

$$V_{max} = C_V \cdot \frac{S}{T}; \quad C_V = 1.6284$$

$$T = \frac{\alpha^\circ}{360N}$$

Steps in calculating displacement table for a specific cam

(1) Select P and check C_A and C_D;

(2) Calculate S_A and S_D.

(3) Multiply Table II-A by S_A.

(4) Multiply Table II-D by S_D and add S_A to every item.

Velocity V_{max}

During Acceleration		
0	0.0000	180
6	0.0458	174
12	0.0916	168
18	0.1375	162
24	0.1833	156
30	0.2291	150
36	0.2749	144
42	0.3208	138
48	0.3666	132
54	0.4123	126
60	0.4576	120
66	0.5024	114
72	0.5465	108
78	0.5895	102
84	0.6314	96
90	0.6718	90
		During Deceleration

During Acceleration		
96	0.7107	84
102	0.7478	78
108	0.7828	72
114	0.8158	66
120	0.8465	60
126	0.8747	54
132	0.9003	48
138	0.9232	42
144	0.9433	36
150	0.9604	30
156	0.9746	24
162	0.9857	18
168	0.9936	12
174	0.9984	6
180	1.0000	0
		During Deceleration

MODIFIED SINE, DIAGRAM II; c = .25

II-A ACCELERATION PERIOD

0	000 000								
1	000 034	37	047 292	73	183 852	109	404 012	145	689 055
2	000 138	38	049 883	74	188 894	110	411 170	146	697 629
3	000 311	39	052 543	75	194 002	111	418 379	147	706 231
4	000 553	40	055 272	76	199 174	112	425 638	148	714 858
5	000 864	41	058 070	77	204 411	113	432 946	149	723 510
6	001 244	42	060 937	78	209 713	114	440 302	150	732 187
7	001 693	43	063 874	79	215 078	115	447 707	151	740 888
8	002 211	44	066 879	80	220 507	116	455 158	152	749 612
9	002 798	45	069 953	81	226 000	117	462 657	153	758 357
10	003 454	46	073 097	82	231 556	118	470 201	154	767 124
11	004 179	47	076 310	83	237 174	119	477 792	155	775 910
12	004 974	48	079 592	84	242 855	120	485 427	156	784 718
13	005 837	49	082 943	85	248 598	121	493 106	157	793 544
14	006 769	50	086 362	86	254 403	122	500 830	158	802 388
15	007 772	51	089 851	87	260 269	123	508 597	159	811 251
16	008 843	52	093 409	88	266 196	124	516 407	160	820 130
17	009 983	53	097 036	89	272 184	125	524 258	161	829 025
18	011 193	54	100 731	90	278 232	126	532 150	162	837 935
19	012 471	55	104 495	91	284 340	127	540 083	163	846 859
20	013 818	56	108 328	92	290 507	128	548 056	164	855 797
21	015 234	57	112 229	93	296 733	129	556 068	165	864 748
22	016 719	58	116 198	94	303 018	130	564 120	166	873 710
23	018 274	59	120 236	95	309 361	131	572 209	167	882 684
24	019 898	60	124 342	96	315 762	132	580 335	168	891 668
25	021 591	61	128 517	97	322 220	133	588 498	169	900 662
26	023 353	62	132 759	98	328 735	134	596 697	170	909 664
27	025 183	63	137 069	99	335 306	135	604 931	171	918 675
28	027 083	64	141 446	100	341 933	136	613 199	172	927 693
29	029 052	65	145 891	101	348 616	137	621 501	173	936 717
30	031 090	66	150 403	102	355 354	138	629 837	174	945 747
31	033 198	67	154 982	103	362 145	139	638 205	175	954 782
32	035 374	68	159 628	104	368 991	140	646 605	176	963 821
33	037 619	69	164 340	105	375 890	141	655 037	177	972 863
34	039 934	70	169 119	106	382 842	142	663 496	178	981 907
35	042 317	71	173 964	107	389 847	143	671 987	179	990 953
36	044 770	72	178 875	108	396 904	144	680 507	180	1.000 000

MODIFIED SINE, DIAGRAM II; c = .25

II-D DECELERATION PERIOD

0	000 000								
1	009 047	37	328 013	73	610 153	109	826 036	145	957 683
2	018 093	38	336 504	74	617 158	110	830 881	146	960 066
3	027 137	39	344 963	75	624 110	111	835 660	147	962 381
4	036 179	40	353 395	76	631 009	112	840 372	148	964 626
5	045 218	41	361 795	77	637 855	113	845 018	149	966 802
6	054 253	42	370 163	78	644 646	114	849 597	150	968 910
7	063 283	43	378 499	79	651 384	115	854 109	151	970 948
8	072 307	44	386 801	80	658 067	116	858 554	152	972 917
9	081 325	45	395 069	81	664 694	117	862 931	153	974 817
10	090 336	46	403 303	82	671 265	118	867 241	154	976 647
11	099 338	47	411 502	83	677 780	119	871 483	155	978 409
12	108 332	48	419 665	84	684 238	120	875 658	156	980 102
13	117 316	49	427 791	85	690 639	121	879 764	157	981 726
14	126 290	50	435 880	86	696 982	122	883 802	158	983 281
15	135 252	51	443 932	87	703 267	123	887 771	159	984 766
16	144 203	52	451 944	88	709 493	124	891 672	160	986 182
17	153 141	53	459 917	89	715 660	125	895 505	161	987 529
18	162 065	54	467 850	90	721 768	126	899 269	162	988 807
19	170 975	55	475 742	91	727 816	127	902 964	163	990 017
20	179 870	56	483 593	92	733 804	128	906 591	164	991 157
21	188 749	57	491 403	93	739 731	129	910 149	165	992 228
22	197 612	58	499 170	94	745 597	130	913 638	166	993 231
23	206 456	59	506 894	95	751 402	131	917 057	167	994 163
24	215 282	60	514 573	96	757 145	132	920 408	168	995 026
25	224 090	61	522 208	97	762 826	133	923 690	169	995 821
26	232 876	62	529 799	98	768 444	134	926 903	170	996 546
27	241 643	63	537 343	99	774 000	135	930 047	171	997 202
28	250 388	64	544 842	100	779 493	136	933 121	172	997 789
29	259 112	65	552 293	101	784 922	137	936 126	173	998 307
30	267 813	66	559 698	102	790 287	138	939 063	174	998 756
31	276 490	67	567 054	103	795 589	139	941 930	175	999 136
32	285 142	68	574 362	104	800 826	140	944 728	176	999 447
33	293 769	69	581 621	105	805 998	141	947 457	177	999 689
34	302 371	70	588 830	106	811 106	142	950 117	178	999 862
35	310 945	71	595 988	107	816 148	143	952 708	179	999 966
36	319 403	72	603 096	108	821 125	144	955 230	180	1.000 000

DIAGRAM III

Modified Sine; $c = 0.25$

The stroke starts from a dwell but is reversed without a dwell.

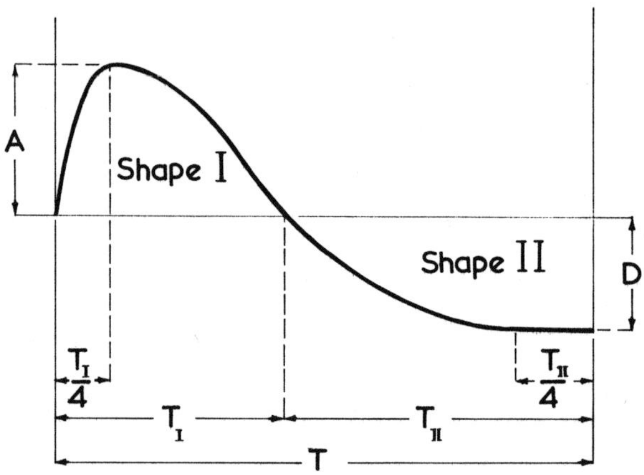

Fig. III - 9

Equations:

$$T_I = \frac{1}{1+P} T; \qquad S_{IA} \text{ or } S_{ID} = \frac{1}{1+1.0805\,P} S;$$

$$\alpha_I = \frac{1}{1+P} \alpha;$$

$$T_{II} = \frac{P}{1+P} T; \qquad S_{IID} \text{ or } S_{IIA} = \frac{1.0805\,P}{1+1.0805\,P} S;$$

$$\alpha_{II} = \frac{P}{1+P} \alpha;$$

$$A_I \text{ or } D_I = C_I \cdot \frac{S}{T^2}; \qquad C_I = 5.5280 \frac{1+P}{2} \cdot \frac{1+P}{1+1.0805\,P};$$

$$D_{II} \text{ or } A_{II} = C_{II} \cdot \frac{S}{T^2}; \qquad C_{II} = 4.8376\, \frac{1+P}{2P} \cdot \frac{1+P}{1+1.0805\,P};$$

$$\frac{C_I}{C_{II}} = 1.1427\,P$$

$$V_{max} = C_V \cdot \frac{S}{T}; \qquad C_V = 1.7596 \cdot \frac{1+P}{1+1.0805\,P}.$$

$$T = \frac{\alpha^\circ}{360N}$$

Tables for the displacements are the same as for Diagrams I and II, according to the subscripts I-A; I-D; II-A; II-D.

The diagram is a combination of two shapes: one—from Diagram I and the other—from Diagram II.

Either one of them may represent the acceleration period.

In this case the ratio P has a different meaning than in cases of symmetrical Diagrams I and II.

$$P = \frac{\text{Period occupied by the Shape II (acceleration or deceleration)}}{\text{Period occupied by the Shape I}} =$$

$$= \frac{T_{II}}{T_I}$$

Displacements S, time T, acceleration A and deceleration D have subscripts different than used for symmetrical Diagrams I and II.

The subscripts I and II indicate that the item belongs to a certain Shape (I or II).

Displacements S have additional subscripts—A and D—designating that shapes I or II are used for acceleration or for deceleration periods; these subscripts also indicate to which unity table the displacement should be applied.

The diagram, as shown above, has S_{IA} and S_{IID}; so S_{IA} should be used with the unity displacement table I-A and S_{IID} with the unity displacement table II-D.

The tables for velocities are used in a similar way.

DIAGRAM I

Sine; b = c = 0. 5

The stroke is performed between dwells.

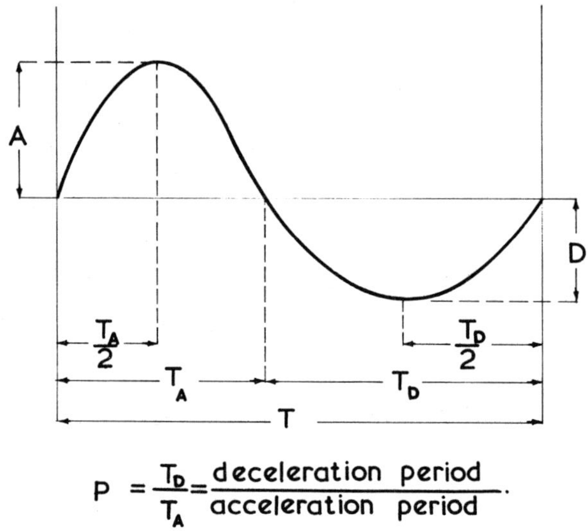

Fig. III - 10

Equations:

$$T_A = \frac{1}{1+P} \cdot T; \quad S_A = \frac{1}{1+P} S; \quad \alpha_A = \frac{1}{1+P} \alpha;$$

$$T_D = \frac{P}{1+P} \cdot T; \quad S_D = \frac{P}{1+P} \quad \alpha_D = \frac{P}{1+P} \alpha;$$

$$A = C_A \cdot \frac{S}{T^2}; \quad C_A = 6.2832 \frac{1+P}{2};$$

$$\frac{C_A}{C_D} = P.$$

$$D = C_D \cdot \frac{S}{T^2}; \quad C_D = 6.2832 \frac{1+P}{2P};$$

$$V_{max} = C_V \cdot \frac{S}{T}; \quad C_V = 2.$$

$$T = \frac{\alpha^\circ}{360N};$$

Steps in calculating displacement table for a specific cam

(1) Select P and check C_A and C_D;

(2) Calculate S_A and S_D;

(3) Multiply Table I-A by S_A.

(4) Multiply Table I-D by S_D and add S_A to every item.

Velocity
V_{max}

During Acceleration			During Acceleration		
0	0.0000	180	96	0.5523	84
6	0.0027	174	102	0.6040	78
12	0.0109	168	108	0.6545	72
18	0.0244	162	114	0.7034	66
24	0.0433	156	120	0.7500	60
30	0.0670	150			
36	0.0955	144	126	0.7939	54
42	0.1284	138	132	0.8346	48
48	0.1654	132	132	0.8716	42
54	0.2061	126	144	0.9045	36
60	0.2500	120	150	0.9330	30
66	0.2966	114	156	0.9568	24
72	0.3455	108	162	0.9755	18
78	0.3960	102	168	0.9891	12
84	0.4477	96	174	0.9973	6
90	0.5000	90	180	1.0000	0
		During Deceleration			During Deceleration

SINE, DIAGRAM I; b = c = .5

I-A ACCELERATION PERIOD

0	000 000								
1	000 000	37	013 992	73	101 154	109	304 588	145	622 980
2	000 002	38	015 140	74	105 132	110	311 998	146	633 114
3	000 008	39	016 348	75	109 203	111	319 499	147	643 303
4	000 018	40	017 617	76	113 367	112	327 090	148	653 544
5	000 035	41	018 948	77	117 625	113	334 771	149	663 836
6	000 061	42	020 343	78	121 979	114	342 543	150	674 178
7	000 097	43	021 802	79	126 427	115	350 402	151	684 569
8	000 144	44	023 328	80	130 970	116	358 349	152	695 007
9	000 205	45	024 921	81	135 608	117	366 384	153	705 490
10	000 282	46	026 582	82	140 343	118	374 505	154	716 018
11	000 375	47	028 313	83	145 174	119	382 711	155	726 588
12	000 486	48	030 116	84	150 101	120	391 002	156	737 198
13	000 618	49	031 990	85	155 123	121	399 377	157	747 848
14	000 772	50	033 938	86	160 243	122	407 836	158	758 537
15	000 949	51	035 960	87	165 459	123	416 377	159	769 261
16	001 151	52	038 057	88	170 773	124	424 998	160	780 020
17	001 380	53	040 231	89	176 183	125	433 700	161	790 813
18	001 637	54	042 482	90	181 690	126	442 482	162	801 637
19	001 924	55	044 811	91	187 294	127	451 342	163	812 491
20	002 243	56	047 220	92	192 995	128	460 279	164	823 373
21	002 576	57	049 710	93	198 793	129	469 293	165	834 282
22	002 982	58	052 281	94	204 688	130	478 383	166	845 216
23	003 404	59	054 933	95	210 679	131	487 547	167	856 174
24	003 865	60	057 669	96	216 767	132	496 783	168	867 153
25	004 365	61	060 489	97	222 952	133	506 092	169	878 152
26	004 906	62	063 394	98	229 232	134	515 471	170	889 170
27	005 490	63	066 384	99	235 609	135	524 922	171	900 205
28	006 118	64	069 460	100	242 081	136	534 439	172	911 255
29	006 791	65	072 624	101	248 649	137	544 024	173	922 319
30	007 511	66	075 876	102	255 313	138	553 676	174	933 395
31	008 280	67	079 216	103	262 070	139	563 392	175	944 480
32	009 098	68	082 645	104	268 923	140	573 172	176	955 574
33	009 969	69	086 165	105	275 870	141	583 014	177	966 674
34	010 892	70	089 775	106	282 910	142	592 918	178	977 780
35	011 869	71	093 476	107	290 043	143	602 881	179	988 889
36	012 902	72	097 269	108	297 269	144	612 902	180	1.000 000

SINE, DIAGRAM I; b = c = .5

I-D

DECELERATION PERIOD

0	000 000								
1	011 111	37	397 119	73	709 957	109	906 524	145	988 131
2	022 220	38	407 082	74	717 090	110	910 225	146	989 108
3	033 326	39	416 986	75	724 130	111	913 835	147	990 031
4	044 426	40	426 828	76	731 077	112	917 355	148	990 902
5	055 520	41	436 608	77	737 930	113	920 784	149	991 720
6	066 605	42	446 324	78	744 687	114	924 124	150	992 489
7	077 681	43	455 976	79	751 351	115	927 376	151	993 209
8	088 745	44	465 561	80	757 919	116	930 540	152	993 882
9	099 795	45	475 078	81	764 391	117	933 616	153	994 510
10	110 830	46	484 529	82	770 768	118	936 606	154	995 094
11	121 848	47	493 908	83	777 048	119	939 511	155	995 635
12	132 847	48	503 217	84	783 233	120	942 331	156	996 135
13	143 826	49	512 453	85	789 321	121	945 067	157	996 596
14	154 784	50	521 617	86	795 312	122	947 719	158	997 018
15	165 718	51	530 707	87	801 207	123	950 290	159	997 424
16	176 627	52	539 721	88	807 005	124	952 780	160	997 757
17	187 509	53	548 658	89	812 706	125	955 189	161	998 076
18	198 363	54	557 518	90	818 310	126	957 518	162	998 363
19	209 187	55	566 300	91	823 817	127	959 769	163	998 620
20	219 980	56	575 002	92	829 227	128	961 943	164	998 849
21	230 739	57	583 023	93	834 541	129	964 040	165	999 051
22	241 463	58	592 164	94	839 757	130	966 062	166	999 228
23	252 152	59	600 623	95	844 877	131	968 010	167	999 382
24	262 802	60	608 998	96	849 899	132	969 884	168	999 514
25	273 412	61	617 289	97	854 826	133	971 687	169	999 625
26	283 982	62	625 495	98	859 657	134	973 418	170	999 718
27	294 510	63	633 616	99	864 392	135	975 079	171	999 795
28	304 993	64	641 651	100	869 030	136	976 672	172	999 856
29	315 431	65	649 598	101	873 573	137	978 198	173	999 903
30	325 822	66	657 457	102	878 021	138	979 657	174	999 939
31	336 164	67	665 229	103	882 375	139	981 052	175	999 965
32	346 456	68	672 910	104	886 633	140	982 383	176	999 982
33	356 697	69	680 501	105	890 797	141	983 652	177	999 992
34	366 886	70	688 002	106	894 868	142	984 860	178	999 998
35	377 020	71	695 412	107	898 846	143	986 008	179	1.000 000
36	387 098	72	702 731	108	902 731	144	987 098	180	1.000 000

DIAGRAM II

Sine; $b = c = 0.5$

The movement is reversed without a dwell.

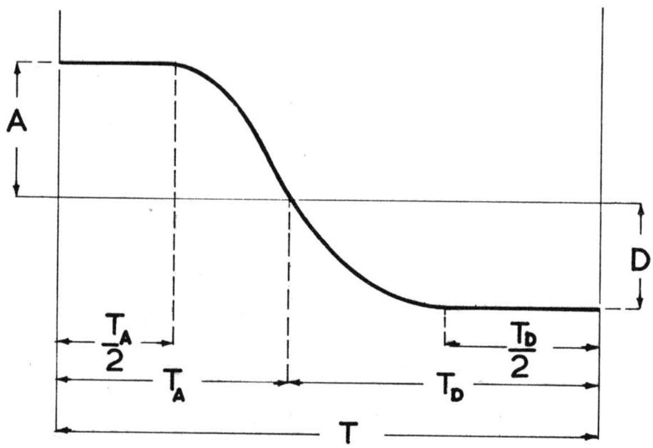

$$P = \frac{T_D}{T_A} = \frac{\text{deceleration period}}{\text{acceleration period}}.$$

Fig. III - 11

Equations:

$$T_A = \frac{1}{1+P}\,T; \qquad S_A = \frac{1}{1+P}\,S; \qquad \alpha_A = \frac{1}{1+P}\,\alpha;$$

$$T_D = \frac{P}{1+P}\,T; \qquad S_D = \frac{P}{1+P}\,S; \qquad \alpha_D = \frac{P}{1+P}\,\alpha;$$

$$A = C_A \cdot \frac{S}{T^2}; \qquad C_A = 4.1988\,\frac{1+P}{2};$$

$$\frac{C_A}{C_D} = P.$$

$$D = C_D \cdot \frac{S}{T^2}; \qquad C_D = 4.1988\,\frac{1+P}{2P};$$

$$V_{max} = C_V \cdot \frac{S}{T}; \quad C_V = 1.7180$$

$$T = \frac{\alpha^\circ}{360 \cdot N};$$

Steps in calculating displacement table for a specific cam

(1) Select P and check C_A and C_D;

(2) Calculate S_A and S_D;

(3) Multiply Table II-A by S_A;

(4) Multiply Table II-D by S_D and add S_A to every item.

Velocity
V_{max}

During Acceleration			During Acceleration		
0	0.0000	180	96	0.6517	84
6	0.0407	174	102	0.6919	78
12	0.0815	168	108	0.7312	72
18	0.1222	162	114	0.7692	66
24	0.1629	156	120	0.8055	60
30	0.2037	150	126	0.8396	54
36	0.2444	144	132	0.8713	48
42	0.2851	138	138	0.9001	42
48	0.3259	132	144	0.9257	36
54	0.3666	126	150	0.9479	30
60	0.4073	120	156	0.9664	24
66	0.4481	114	162	0.9810	18
72	0.4888	108	168	0.9915	12
78	0.5296	102	174	0.9979	6
84	0.5703	96	180	1.0000	0
90	0.6110	90			
		During Deceleration			During Deceleration

SINE, DIAGRAM II: b = c = .5

II-A ACCELERATION PERIOD

0	000 000								
1	000 032	37	044 353	73	172 652	109	384 820	145	673 880
2	000 130	38	046 784	74	177 415	110	391 891	146	682 772
3	000 292	39	049 278	75	182 242	111	399 023	147	691 700
4	000 518	40	051 838	76	187 134	112	406 216	148	700 662
5	000 810	41	054 462	77	192 091	113	413 468	149	709 659
6	001 166	42	057 151	78	197 113	114	420 780	150	718 690
7	001 588	43	059 905	79	202 199	115	428 151	151	727 753
8	002 074	44	062 724	80	207 350	116	435 581	152	736 847
9	002 624	45	065 607	81	212 566	117	443 069	153	745 972
10	003 240	46	068 555	82	217 848	118	450 615	154	755 126
11	003 920	47	071 568	83	223 194	119	458 218	155	764 309
12	004 665	48	074 646	84	228 604	120	465 879	156	773 519
13	005 475	49	077 788	85	234 079	121	473 594	157	782 755
14	006 350	50	080 997	86	239 619	122	481 366	158	792 017
15	007 290	51	084 269	87	245 224	123	489 192	159	801 303
16	008 294	52	087 606	88	250 894	124	497 073	160	810 612
17	009 363	53	091 008	89	256 629	125	505 008	161	819 943
18	010 497	54	094 474	90	262 428	126	512 996	162	829 296
19	011 696	55	098 006	91	268 292	127	521 036	163	838 668
20	012 960	56	101 602	92	274 221	128	529 128	164	848 059
21	014 288	57	105 263	93	280 215	129	537 270	165	857 469
22	015 681	58	108 989	94	286 273	130	545 463	166	866 895
23	017 139	59	112 779	95	292 396	131	553 706	167	876 337
24	018 662	60	116 634	96	298 584	132	561 998	168	885 793
25	020 249	61	120 555	97	304 836	133	570 338	169	895 263
26	021 901	62	124 540	98	311 152	134	578 725	170	904 745
27	023 618	63	128 590	99	317 532	135	587 159	171	914 238
28	025 401	64	132 705	100	323 976	136	595 639	172	923 741
29	027 247	65	136 884	101	330 485	137	604 164	173	933 254
30	029 159	66	141 128	102	337 057	138	612 732	174	942 774
31	031 135	67	145 437	103	343 692	139	621 345	175	952 302
32	033 176	68	149 811	104	350 390	140	630 001	176	961 834
33	035 282	69	154 250	105	357 152	141	638 696	177	971 372
34	037 453	70	158 753	106	363 976	142	647 433	178	980 913
35	039 688	71	163 321	107	370 862	143	656 210	179	990 456
36	041 988	72	167 954	108	377 810	144	665 026	180	1.000 000

SINE, DIAGRAM II; b = c = .5

II-D DECELERATION PERIOD

0	000 000								
1	009 544	37	343 790	73	629 138	109	836 679	145	960 312
2	019 087	38	352 567	74	636 024	110	841 247	146	962 547
3	028 628	39	361 304	75	642 848	111	845 750	147	964 718
4	038 166	40	369 999	76	649 610	112	850 189	148	966 824
5	047 698	41	378 655	77	656 308	113	854 563	149	968 865
6	057 226	42	387 268	78	662 943	114	858 872	150	970 841
7	066 746	43	395 836	79	669 515	115	863 116	151	972 753
8	076 259	44	404 361	80	676 024	116	867 295	152	974 599
9	085 762	45	412 841	81	682 468	117	871 410	153	976 382
10	095 255	46	421 275	82	688 848	118	875 460	154	978 099
11	104 737	47	429 662	83	695 164	119	879 445	155	979 751
12	114 207	48	438 002	84	701 416	120	883 366	156	981 338
13	123 663	49	446 294	85	707 604	121	887 221	157	982 861
14	133 105	50	454 537	86	713 727	122	891 011	158	984 319
15	142 531	51	462 730	87	719 785	123	894 737	159	985 712
16	151 941	52	470 872	88	725 779	124	898 398	160	987 040
17	161 332	53	478 964	89	731 708	125	901 994	161	988 304
18	170 704	54	487 004	90	737 572	126	905 526	162	989 503
19	180 057	55	494 992	91	743 371	127	908 992	163	990 637
20	189 388	56	502 927	92	749 106	128	912 394	164	991 706
21	198 697	57	510 808	93	754 770	129	915 731	165	992 710
22	207 983	58	518 634	94	760 381	130	919 003	166	993 650
23	217 245	59	526 406	95	765 921	131	922 212	167	994 525
24	226 481	60	534 121	96	771 396	132	925 354	168	995 335
25	235 691	61	541 782	97	776 806	133	928 432	169	996 080
26	244 874	62	549 385	98	782 152	134	931 445	170	996 760
27	254 028	63	556 931	99	787 434	135	934 393	171	997 376
28	263 153	64	564 419	100	792 650	136	937 276	172	997 926
29	272 247	65	571 849	101	797 801	137	940 095	173	998 412
30	281 310	66	579 220	102	802 887	138	942 849	174	998 834
31	290 341	67	586 532	103	807 909	139	945 538	175	999 190
32	299 338	68	593 784	104	812 866	140	948 162	176	999 482
33	308 300	69	600 977	105	817 758	141	950 722	177	999 708
34	317 228	70	608 109	106	822 585	142	953 216	178	999 870
35	326 120	71	615 180	107	827 348	143	955 647	179	999 968
36	334 974	72	622 190	108	832 046	144	958 012	180	1.000 000

DIAGRAM III

Sine; b = c = 0.5

The stroke starts from a dwell but is reversed without a dwell.

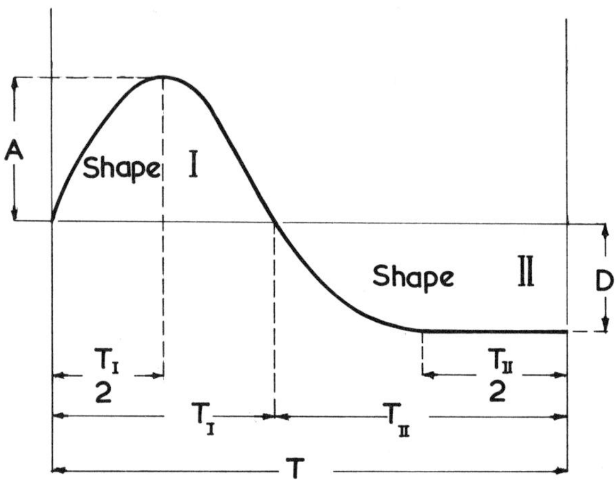

Fig. III - 12

Equations:

$$T_I = \frac{1}{1 + P} T; \qquad S_{IA \text{ or } ID} = \frac{1}{1 + 1.1642 P} \cdot S;$$

$$\alpha_I = \frac{1}{1 + P} \alpha;$$

$$T_{II} = \frac{P}{1 + P} T; \qquad S_{IID \text{ or } IIA} = \frac{1.1642 P}{1 + 1.1642 P} S;$$

$$\alpha_{II} = \frac{P}{1 + P} \alpha;$$

$$A_I \text{ or } D_I = C_I \cdot \frac{S}{T^2}; \qquad C_I = 6.2832 \cdot \frac{1 + P}{2} \cdot \frac{1 + P}{1 + 1.1642 P};$$

$$D_{II} \text{ or } A_{II} = C_{II} \cdot \frac{S}{T^2}; \qquad C_{II} = 4.8881 \frac{1 + P}{2P} \cdot \frac{1 + P}{1 + 1.1642 P};$$

$$\frac{C_I}{C_{II}} = 1.2854 P.$$

$$V_{max} = C_V \cdot \frac{S}{T}; \qquad C_V = 2\,\frac{1 + P}{1 + 1.1642\,P}.$$

$$T = \frac{\alpha^\circ}{360N}.$$

Tables for the displacements are the same as for Diagrams I and II, according to the subscripts I-A; I-D; II-A; II-D.

The diagram is a combination of two shapes: one—from Diagram I and the other—from Diagram II.

Either one of them may represent the acceleration period.

In this case the ratio P has a different meaning than in cases of symmetrical Diagrams I and II.

$$P = \frac{\text{Period occupied by the Shape II (acceleration or deceleration)}}{\text{Period occupied by the Shape I}} =$$

$$= \frac{T_{II}}{T_I}.$$

Displacements S, time T, acceleration A and deceleration D have subscripts different than used for symmetrical Diagrams I and II.

The subscripts I and II indicate that the item belongs to a certain Shape (I or II).

Displacements S have additional subscripts—A and D—designating that shapes I or II are used for acceleration or for deceleration periods; these subscripts also indicate to which unity table the displacement should be applied.

The diagram, as shown above, has S_{IA} and S_{IID}; so S_{IA} should be used with the unity displacement Table I-A and S_{IID}—with the unity displacement Table II-D.

The tables for velocities are used in a similar way.

Development of Additional Diagrams

As it was mentioned before—the search for "ideal" acceleration diagram should be continued, because requirements vary, so no single solution can be the best for all of them. The offered three groups of acceleration diagrams are based on trigonometric functions and their shapes are affected by preselected factors, designated as b or c.

The introduction of the factor P permits to modify all these diagrams, but it permits to use precalculated tables of displacements prepared for certain factors b or c with $P = 1$. Modified Trapezoid Diagram I utilizes period of uniform acceleration to decrease the value of C_A and it is symmetrical, because the variations of magnitude from 0 to max. have the same period and characteristics as the variations from max. to 0; therefore $C_V = 2$.

Modified Sine Diagram I utilizes only quadrants of Sine curve, but the period of changes from 0 to +max. is different than the period of change from +max. to zero, so the diagram of +acceleration is not symmetrical and $C_V < 2$. It is possible to create other groups of Diagrams I, II and III by combinations of peculiarities of Modified Trapezoid and of Modified Sine diagrams.

For instance, some period of Uniform acceleration can be retained to keep C_A lower and at the same time periods of change from 0 to +max. and from +max. to 0 can be different, so the Diagram I is not symmetrical and C_V can be <2.

As an illustration, a new group of diagrams can be developed by the introduction of an additional factor n which has a value greater than unity. Suppose, the factor b is retained for both ends of the whole Diagram I and a factor nb is used for periods when the magnitude of the acceleration changes from +max. to 0 and from 0 to —max., in other words, the factor nb is used for the middle part of the whole Diagram I.

To retain the period of Uniform acceleration, it is necessary to have a positive value of the expression $1.0 - (b + nb) = 1.0 - b(1 + n)$.

If $b = 0.25$ and $n = 2$, Uniform acceleration occupies 0.25 of the acceleration and of the deceleration periods of the Diagram I.

With these values of b and n the Diagram II is exactly the same as in the case of Sine curve accelerat ion, because it has $b = c = 0.5$ and in the new Diagram II $nb = 2 \times 0.25 = 0.5$, or the same value.

Therefore, the new group of Diagrams I, II and III requires displacement tables only for the Diagram I and new equations for the Diagram III; displacement tables and equations for the Diagram II are on hand for Sine Diagram II with $b = c = 0.5$. The new illustrative group is named "Sine and Uniform Acceleration with factors $b = 0.25$; $n = 2$." Its properties are in between the original Modified Trapezoid with $b = 0.25$ and Modified Sine with $c = 0.25$.

Residue vibrational characteristics are about the same as in the original diagrams. Vibrational stresses during the stroke are close to stresses in case of Modified Sine Acceleration.

DIAGRAM I

Sine and Uniform Acceleration with Factors b and n

$b = 0.25; n = 2; nb = 0.5$

The stroke is performed between dwells

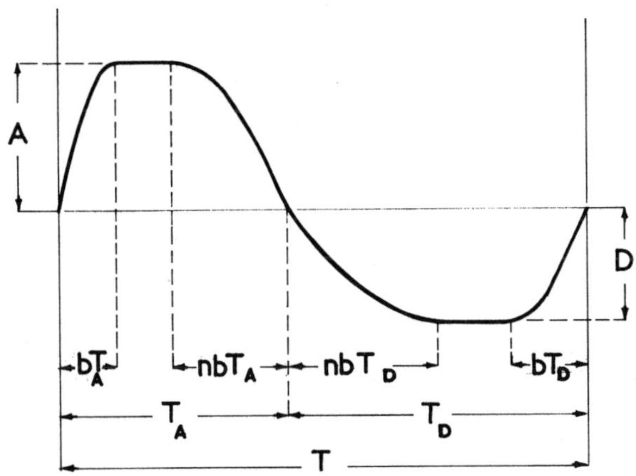

$$P = \frac{T_D}{T_A} = \frac{\text{deceleration period}}{\text{acceleration period}}.$$

Fig. III - 13

$$\frac{T_D}{T_A} = \frac{\text{Deceleration Period}}{\text{Acceleration Period}} = P$$

$$T_A = \frac{1}{1+P} T; \qquad S_A = \frac{1}{1+P} S; \qquad \alpha_A = \frac{1}{1+P} \alpha.$$

$$T_D = \frac{P}{1+P} T; \qquad S_D = \frac{P}{1+P} S; \qquad \alpha_D = \frac{P}{1+P} \alpha.$$

$$A = C_A \frac{S}{T^2}; \qquad C_A = 5.10992 \frac{1+P}{2};$$

$$\frac{C_A}{C_D} = P.$$

$$D = C_D \frac{S}{T^2}; \qquad C_D = 5.10992 \frac{1+P}{2P};$$

$$V_{max} = C_V \frac{S}{T}; \qquad C_V = 1.8588.$$

$$C_A = \frac{(1 + P)2\pi}{\pi - (\pi - 2)2b - (n^2 - 1)\dfrac{\pi^2 - 8}{\pi} b^2}.$$

$$C_V = 2\,\frac{\pi - (\pi - 2)b\,(1 + n)}{\pi - (\pi - 2)2b - (n^2 - 1)\dfrac{\pi^2 - 8}{\pi} b^2}.$$

Steps in calculating displacement table for a specific cam:

(1) Select P;

(2) Calculate S_A and S_D;

(3) Multiply the table I-A by S_A;

(4) Multiply the table I-D by S_D and add S_A.

Velocity V_{max}

During Acceleration		
0	0.0000	180
6	0.0048	174
12	0.0189	168
18	0.0418	162
24	0.0724	156
30	0.1094	150
36	0.1512	144
42	0.1959	138
48	0.2417	132
54	0.2875	126
60	0.3333	120
66	0.3792	114
72	0.4250	108
78	0.4708	102
84	0.5166	96
		During Deceleration

During Acceleration		
90	0.5626	90
96	0.6082	84
102	0.6534	78
108	0.6977	72
114	0.7404	66
120	0.7812	60
126	0.8196	54
132	0.8552	48
138	0.8876	42
144	0.9164	36
150	0.9414	30
156	0.9643	24
162	0.9786	18
168	0.9904	12
174	0.9976	6
180	1.0000	0
		During Deceleration

SINE AND UNIFORM ACCELERATION DIAGRAM; b = .25; n= 2; P = 1

I-A UNITY DISPLACEMENT TABLE FOR THE ACCELERATION PERIOD

0	000 000								
1	000 000	37	021 375	73	131 107	109	342 890	145	648 250
2	000 004	38	023 050	74	135 613	110	350 206	146	656 781
3	000 012	39	024 800	75	140 199	111	357 596	147	667 357
4	000 029	40	026 628	76	144 863	112	365 060	148	676 975
5	000 057	41	028 534	77	149 606	113	372 596	149	686 635
6	000 099	42	030 518	78	154 428	114	380 205	150	696 335
7	000 157	43	032 580	79	159 329	115	387 887	151	706 075
8	000 234	44	034 721	80	164 308	116	395 639	152	715 853
9	000 333	45	036 941	81	169 367	117	403 463	153	725 669
10	000 456	46	039 239	82	174 504	118	411 357	154	735 520
11	000 606	47	041 617	83	179 720	119	419 320	155	745 405
12	000 786	48	044 073	84	185 015	120	427 353	156	755 324
13	000 998	49	046 608	85	190 389	121	435 453	157	765 275
14	001 244	50	049 222	86	195 842	122	443 622	158	775 257
15	001 527	51	051 915	87	201 374	123	451 857	159	785 268
16	001 850	52	054 686	88	206 984	124	460 158	160	795 308
17	002 215	53	057 537	89	212 673	125	468 525	161	805 375
18	002 624	54	060 466	90	218 442	126	476 956	162	815 467
19	003 078	55	063 474	91	224 289	127	485 452	163	825 584
20	003 582	56	066 561	92	230 215	128	494 010	164	835 723
21	004 136	57	069 727	93	236 219	129	502 630	165	845 885
22	004 743	58	072 972	94	242 303	130	511 311	166	856 066
23	005 405	59	076 296	95	248 465	131	520 053	167	866 267
24	006 123	60	079 698	96	254 706	132	528 855	168	876 486
25	006 900	61	083 180	97	261 025	133	537 715	169	886 721
26	007 738	62	086 740	98	267 422	134	546 632	170	896 971
27	008 637	63	090 379	99	273 897	135	555 607	171	907 234
28	009 601	64	094 097	100	280 451	136	564 637	172	917 511
29	010 629	65	097 894	101	287 082	137	573 722	173	927 798
30	011 725	66	101 769	102	293 790	138	582 861	174	938 095
31	012 889	67	105 724	103	300 576	139	592 052	175	948 400
32	014 122	68	109 757	104	307 438	140	601 295	176	958 711
33	015 427	69	113 869	105	314 377	141	610 589	177	969 029
34	016 803	70	118 060	106	321 392	142	619 933	178	979 350
35	018 252	71	122 330	107	328 483	143	628 325	179	989 674
36	019 776	72	126 679	108	335 649	144	638 764	180	1.000 000

SINE AND UNIFORM ACCELERATION DIAGRAM I; b = .25; n = 2; P = 1

I-D UNITY DISPLACEMENT TABLE FOR THE DECELERATION PERIOD

0	000 000								
1	010 326	37	371 675	73	671 517	109	877 670	145	981 748
2	020 650	38	380 067	74	678 608	110	881 940	146	983 197
3	030 971	39	389 411	75	685 623	111	886 130	147	984 573
4	041 288	40	398 705	76	692 562	112	890 243	148	985 877
5	051 600	41	407 948	77	699 424	113	894 276	149	987 111
6	061 905	42	417 139	78	706 210	114	898 231	150	988 274
7	072 202	43	426 278	79	712 918	115	902 106	151	989 370
8	082 489	44	435 363	80	719 549	116	905 903	152	990 399
9	092 765	45	444 393	81	726 102	117	909 621	153	991 363
10	103 029	46	453 368	82	732 578	118	913 260	154	992 262
11	113 279	47	462 285	83	738 975	119	916 820	155	993 100
12	123 514	48	471 145	84	745 294	120	920 302	156	993 877
13	133 733	49	479 947	85	751 535	121	923 704	157	994 595
14	143 933	50	488 689	86	757 697	122	927 028	158	995 257
15	154 115	51	497 370	87	763 781	123	930 273	159	995 864
16	164 277	52	505 990	88	769 785	124	933 439	160	996 418
17	174 416	53	514 548	89	775 711	125	936 526	161	996 922
18	184 533	54	523 043	90	781 558	126	939 534	162	997 377
19	194 625	55	531 475	91	787 326	127	942 463	163	997 785
20	204 692	56	539 841	92	793 016	128	945 314	164	998 150
21	214 731	57	548 143	93	798 626	129	948 085	165	998 473
22	224 743	58	556 378	94	804 158	130	950 778	166	998 756
23	234 725	59	564 547	95	809 611	131	953 392	167	999 002
24	244 676	60	572 648	96	814 985	132	955 927	168	999 214
25	254 595	61	580 680	97	820 280	133	958 383	169	999 394
26	264 480	62	588 643	98	825 496	134	960 761	170	999 544
27	274 331	63	596 537	99	830 633	135	963 059	171	999 667
28	284 146	64	604 360	100	835 692	136	965 279	172	999 766
29	293 925	65	612 113	101	840 671	137	967 420	173	999 843
30	303 665	66	619 794	102	845 572	138	969 482	174	999 901
31	313 365	67	627 404	103	850 394	139	971 466	175	999 943
32	323 025	68	634 940	104	855 137	140	973 371	176	999 971
33	332 643	69	642 404	105	859 801	141	975 199	177	999 988
34	343 219	70	649 794	106	864 387	142	976 950	178	999 996
35	351 750	71	657 110	107	868 893	143	978 625	179	999 999
36	361 236	72	664 351	108	873 321	144	980 223	180	1.000 000

DIAGRAM III

Sine and Uniform Acceleration with Factors b and n

$b = 0.25; n = 2; nb = 0.5$

The stroke starts from a dwell, but is reversed without a dwell

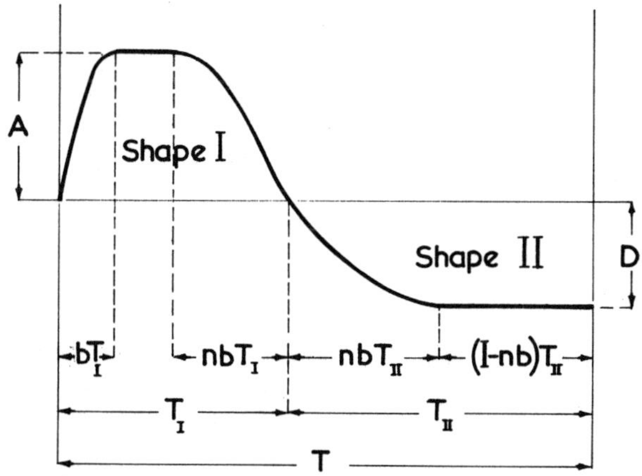

Fig. III - 14

The diagram is a combination of two shapes: one-from Diagram I and the other from Diagram II (Sine Acc).

Either one of them may represent the acceleration period.

In this case the ratio P has a different meaning than in cases of symmetrical Diagrams I and II.

$$P = \frac{T_{II}}{T_I} = \frac{\text{period occupied by the shape II (acceleration or deceleration)}}{\text{period occupied by shape I}}$$

Displacements S, time T, acceleration A and deceleration D have subscripts different than used for symmetrical Diagrams I and II.

The subscripts I and II indicate that the item belongs to a certain shape (I or II).

Displacements S have additional subscripts-A and D-designating that shapes I or II are used for acceleration or for deceleration periods; these additional subscripts also indicate to which Unity table the displacement should be applied.

The diagram, as shown above, has S_{IA} and S_{IID}; so S_{IA} should be used with the Unity displacement table I-A and S_{IID} with the Unity displacement table II-D.

The tables for velocity are used in a similar way.

$$T_I = \frac{1}{1+P} T; \qquad S_{IA} \text{ or } S_{ID} = \frac{1}{1 + 1.08188\ P} S;$$

$$\alpha_I = \frac{1}{1+P} \alpha.$$

$$T_{II} = \frac{P}{1+P} T; \qquad S_{IIA} \text{ or } S_{IID} = \frac{1.08188\ P}{1 + 1.08188\ P} S;$$

$$\alpha_{II} = \frac{P}{1+P} \alpha.$$

$$A_I \text{ or } D_I = C_I \frac{S}{T^2}; \qquad C_I = 5.10992 \frac{(1+P)^2}{2} \cdot \frac{1}{1 + 108188\ P};$$

$$A_{II} \text{ or } D_{II} = C_{II} \frac{S}{T^2}; \qquad C_{II} = 4.54264 \frac{(1+P)^2}{2P} \cdot \frac{1}{1 + 108188\ P};$$

$$\frac{C_I}{C_{II}} = 1.1249\ P.$$

$$V_{max} = C_V \cdot \frac{S}{T}; \qquad C_V = 1.8586 \frac{1+P}{1 + 1.08188\ P}.$$

$$C_I = \frac{(1+P)^2\, 2\pi}{\left[\pi - (\pi - 2)2b - (n^2 - 1)\dfrac{\pi^2 - 8}{\pi} b^2\right] + P \dfrac{\pi - (\pi - 2)b\,(1+n)}{\pi - (\pi - 2)\ bn}\left[\pi - \dfrac{\pi^2 - 8}{\pi} b^2 n^2\right]}$$

$$C_{II} = \frac{C_I}{P} \cdot \frac{\pi - (\pi - 2)b\,(1+n)}{\pi - (\pi - 2)\ bn}.$$

$$C_V = \frac{C_I}{1+P} \cdot \frac{\pi - (\pi - 2)b\,(1+n)}{\pi}$$

Displacement tables for S_{IA} and S_{ID} are the tables I-A and I-D for Diagram I of Sine and Uniform Acceleration with $b = 0.25$ and $\underline{n = 2.}$

Displacement tables for S_{IIA} and S_{IID} are the tables II-A and II-D for Diagram II of Sine with $b = c = 0.5$.

CHAPTER IV

EXAMPLES

It is difficult to offer even a few definite rules for designing cams, which would be good for any conditions. The designer's decisions are affected by numerous factors, as the length of the stroke and the available working period; requirements of a precision and of the absence of back-lash; the magnitude of masses to be moved by the cam, and the work load; a protection of the machine from damaging and, therefore, expensive jams; the available space in the machine and convenient positions of cam shafts; the difference in cost of a simple and a more complicated, though better, cam; and so on, ad infinitum.

Therefore, a few, more or less, typical problems are presented; in the description of requirements some points are brought up to explain the selected solution without a claim of being the best possible one.

In addition, a few rather unusual problems are described, which were solved in practice, and also an analysis of the influence of constant velocity period during the stroke.

The solutions are limited to the determination of dynamic forces and of the data necessary for the preparation of instructions for a machine shop.

When acceleration diagrams are drawn for cams, using a spring to hold the cam follower against the cam surface, the following rule is recommended:

> if the inertial or dynamic force presses the follower against the cam surface, the part of the acceleration diagram producing this pressure is shown above the neutral line;
>
> when there is a tendency for the cam follower to leave the cam surface, the acceleration diagram during that period is shown below the neutral line.

EXAMPLE I

The cam should produce a stroke $S = 2''$ during $\alpha = 90°$; next is a 90° dwell; the return stroke is done during 72°, so the remaining dwell is 108°. A grooved face cam is selected so no spring is required. Speed of the cam 180 RPM, so N = 3 RPS. Work load is negligible—to push a light part into some position in the machine.

The movements are of Type I; the factor P = 1, because there are no special requirements how the displacement changes during the stroke. Modified Trapezoid I with b = 0.25 is selected. The original stroke (α = 90°) is made during the rise (when the cam follower moves away from the center of the cam).

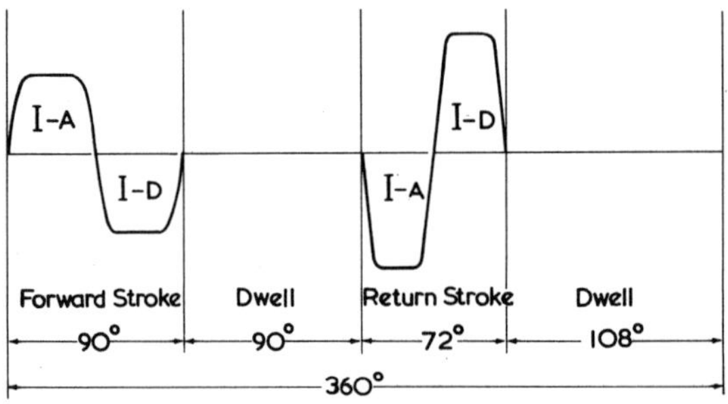

Fig. IV - 1

For both strokes C_A = 4.89 and C_V = 2.0, though the latter is not important.

Forward stroke: maximum acceleration

$$A_{max} = C_A S\left(\frac{360}{\alpha}N\right)^2 = 4.89 \cdot 2(4.3)^2 = 1408 \text{ inch/sec.}^2$$

In terms of gravity acceleration

$$\frac{A_{max}}{g} = \frac{1408}{386} = 3.65$$

so the dynamic force $F = \text{Weight} \cdot \frac{A_{max}}{g} = 3.65 \cdot W$ lb.

Displacements during the acceleration and the deceleration period = 1″; therefore, the Tables I-A and I-D are multiplied by 1″.

Each division in each table $= \frac{\alpha°}{2 \cdot 180} = \frac{90°}{360} = 0.25° = 15'$ of the arc.

Return stroke:

$$A_{max} = 4.89 \cdot 2\left(\frac{360}{72} \cdot 3\right)^2 = 2200 \text{ inch/sec.}^2$$

$$\frac{A_{max}}{g} = 5.7; \qquad F_{max} = 5.7 \cdot W \text{ lb}$$

Both displacement tables—I-A and I-D—are multiplied by 1″. Each division in both tables $= \frac{72°}{360} = 0.2° = 12'$ of the arc.

Remarks: maximum acceleration during the forward stroke is quite high (3.65 g). If a part only pushed into the machine, there should be either a stop or some frictional resistance to prevent the part from traveling away from the pusher; the selection of the Modified Trapezoid Diagram is justified because in case of the Modified Sine Diagram, the acceleration would be about 13% higher or

$$3.65g \cdot \frac{5.53}{4.89} = 4.13 \text{ g.}$$

maybe, an increase of the working period of the working forward stroke is possible; for instance, an increase from 90° to 108° or by 20% would decrease the maximum acceleration to 3.65 g ÷ 1.44 = 2.53 g.

If the cam performs well at 180 RPM, an increase of the working period may permit 20% increase of speed, up to 180 × 1.2 = 216 RPM.

EXAMPLE II

The same problem, but a flat cam is selected with a spring holding the cam follower against the cam surface. The pressure of the spring should be greater than the dynamic force during the deceleration period of the forward stroke and during acceleration period of the return stroke when the cam follower moves toward the center of the cam. The latter is greater; so an attempt should be made to equalize working periods of both strokes.

Suppose, it is possible, so both working periods are equal 90°. Then the pressure on the cam during the acceleration period of the forward stroke will be greater due to the pressure of the spring. In this case, it is proper to have $P \neq 1$.

Select $P = 1.5$, so the deceleration period of the forward stroke is 50% longer than the acceleration period. During the return stroke, the acceleration period should be longer than the decleration period and $P - 2/3$.

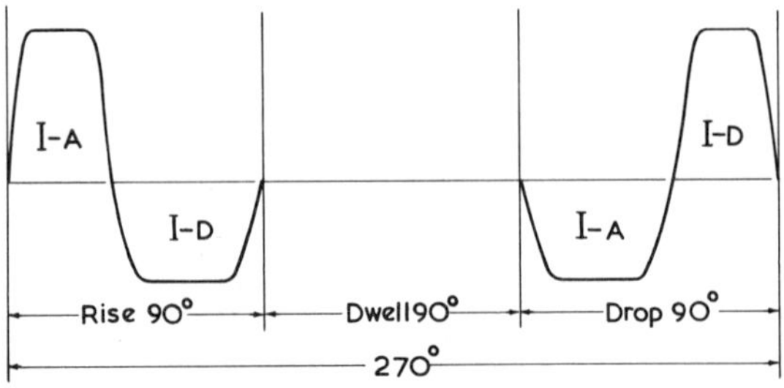

Fig. IV - 2

Forward stroke:

$$P = \frac{3}{2};\ \alpha_A = \frac{1}{1 + P}\alpha = 36°;\ \alpha_D{}° = \alpha° - \alpha_A{}° = 54°.$$

Strokes

$$S_A = \frac{1}{1 + P}S = \frac{2''}{2.5} = 0.8'';\ S_D = \frac{P}{1 + P}S = \frac{1.5}{2.5} \cdot 2'' = 1.2''$$

Coefficients

$$C_A = 4.888\,\frac{1 + P}{2} = 4.888\,\frac{2.5}{2} = 6.11; \qquad C_D = 4.888\,\frac{1 + P}{2P} = 4.073.$$

Maximum acceleration

$$\frac{A_{max}}{g} = \frac{6.11 \cdot 5''}{386}\left(\frac{360}{90} \cdot 3\right)^2 = 4.56;$$

$$\frac{D_{max}}{g} = \frac{4.073 \cdot 2''}{386}\left(\frac{360}{90} \cdot 3\right)^2 = 3.04$$

The displacement table I-A is multiplied by $S_A = 0.8''$ and one division of the table is

$$\frac{\alpha_A{}°}{180} = \frac{36°}{180} = 0.2° = 12' \text{ of the arc.}$$

The displacement table I-D is multiplied by $S_D = 1.2''$ and one division of the table is $\frac{54^\circ}{180} = 0.3^\circ = 18'$ of the arc.

Return stroke: the figures are the same, but the data with the subscript A is applied to the deceleration and with the subscript D is applied to the acceleration, because $P = 2/3$ instead of $P = 3/2$ (see calculation in the Case III).

Remarks: during the forward stroke, the spring should provide the force greater than 3.04. Weight during the deceleration period; the pressure on the cam surface by the cam follower will be greater, it is 4.56 . W during the acceleration period + the varying tension of the spring.

EXAMPLE III

The same problem with a flat cam with a spring holding the cam follower against the cam surface. The part which should be pushed into the machine (or pushed out) may not enter into the place assigned for it and this fact should not cause any damage either to the part or to the plunger with the cam follower system.

Two solutions are possible:

(1) The plunger acting on the part has a precompressed spring between it and the cam follower, so the plunger can press only with a definite force developed by the spring subjected to a further compression;

(2) The forward or working stroke of the plunger is produced by a drop and not by a rise on the cam; therefore, the work of pushing the part is done by the spring which holds the cam follower on the cam surface, so the purpose of the cam is only to control the movement.

The designer selected the second solution.

It means the spring should provide the force necessary to push the part and also to accelerate the cam follower system. In this case, it may be desirable to have a lower acceleration during the forward or working stroke, when the cam follower moves toward the center of the cam. For this purpose, it is necessary to increase the time for the acceleration period, so the factor P should be smaller than unity.

When the acceleration diagram is drawn, it should be remembered that during the acceleration period, the cam follower tends to leave the cam surface; therefore, the acceleration part of the diagram during the working period should be below the neutral line.

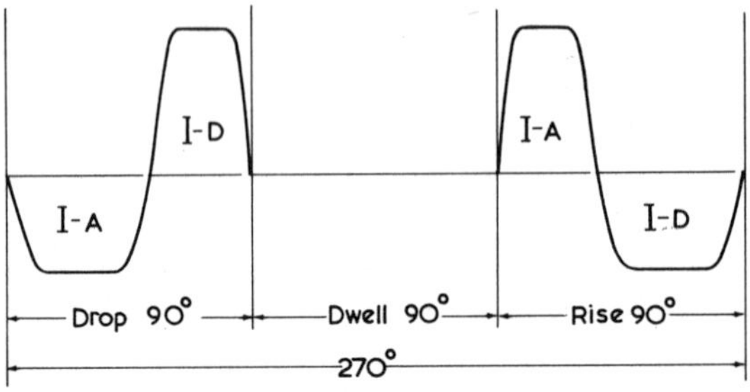

Fig. IV - 3

Forward stroke:

$$P = \frac{2}{3}; \alpha_A^\circ = \frac{1}{1 + P} \alpha^\circ = \frac{3}{5} \cdot 90^\circ = 54^\circ; \qquad \alpha_D = \frac{P}{1 + P} \alpha = \frac{2}{5} \cdot 90^\circ = 36^\circ.$$

Strokes

$$S_A = \frac{S}{1 + P} = \frac{2.3''}{5} = 1.2''; \qquad S_D = S - S_A = 2'' - 1.2'' = 0.8''$$

Coefficients

$$C_A = 4.888 \frac{1 + P}{2} = 4.888 \cdot \frac{5}{6} = 4.073$$

$$C_D = 4.888 \frac{1 + P}{2P} = 4.888 \frac{5}{4} = 6.11$$

The displacement table I-A is multiplied by $S_A = 1.2''$; one division of the table $= \frac{54^\circ}{180^\circ} = 0.3^\circ = 18'$ of the arc.
The displacement table I-D is multiplied by $S_D = 0.8''$; one division of the table $= \frac{36^\circ}{180} = 0.2^\circ = 12'$ of the arc.
During the return stroke, the cam follower rises, so a possibility of separation from the cam surface exists during the deceleration; therefore, it may be advisable to use $P = 1\frac{1}{2}$ for calculation of C_A and C_D.

EXAMPLE IV

A cam should make a stroke $S = 2.5''$ and return to the initial position immediately—without a dwell between strokes. Each stroke has the working period $\alpha° = 110°$; Speed 120 RPM or N = 2 RPS. The cam moves a heavy load; to decrease the torque on the cam shaft, Modified Sine diagram with b = 0.25 is selected.

A flat cam with a spring, holding the cam follower against the cam surface, is selected, due to a convenient position of the shaft for it and a possibility to have the cam follower in a yoke—supported on both sides.

The dwell is selected to be on the smaller radius of the cam, so the first stroke after a dwell is during a rise (increase of the radius) on the cam.

The cam follower tends to leave the cam surface during the deceleration period of the first stroke and also during the acceleration period of the second stroke, which is produced during the drop (decrease of the radius) on the cam.

It is desirable to have smaller dynamic forces during these two periods—to decrease the spring pressure required to overcome them. Select—during the first stroke $P = 3/2$ and also during the return stroke $P = 3/2$.

The movement is of the Type III and, as it was explained before,

$$P = \frac{\text{Period occupied by the Shape II}}{\text{Period occupied by the Shape I}};$$

therefore, the value of P is the same for both strokes.

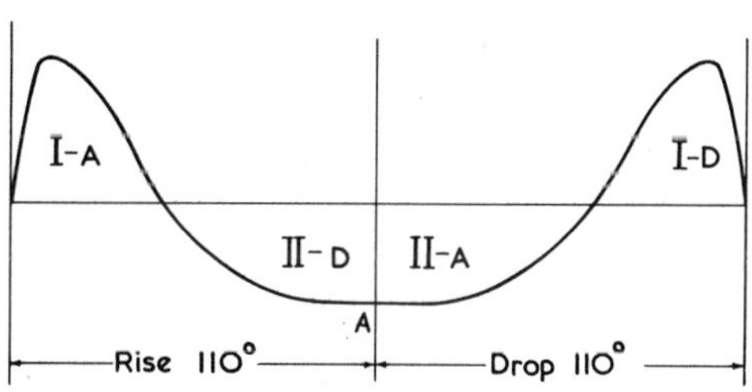

Fig. IV - 4

Forward first stroke:

$$P = \frac{3}{2}; \quad \alpha_{IA} = \frac{1}{1+P}\alpha = \frac{110°}{2.5} = 44°;$$

$$\alpha_{IID} = \frac{P}{1+P}\alpha = \frac{1.5}{2.5} \cdot 110° = 66°.$$

Strokes:

$$S_{IA} = \frac{S}{1 + 1.0805P} = \frac{2.5}{2.62075} = 0.9539'';$$

$$S_{IID} = \frac{1.0805P}{1 + 1.0805P} S = \frac{1.62075}{2.62075} \cdot 2.5'' = 1.5461''$$

Coefficients:

$$C_I = 5.528 \frac{(1 + P)^2}{2(1 + 1.0805P)} = 6.592;$$

$$C_{II} = 4.8376 \frac{(1 + P)^2}{2P(1 + 1.0805P)} = 3.846$$

Max. accel.:

$$\frac{A}{g} = \frac{6.592 \cdot 2.5''}{386} \left(\frac{360}{110} \cdot 2\right)^2 = 1.83;$$

$$\frac{D}{g} = \frac{3.846 \cdot 2.5''}{386} \left(\frac{360}{110} \cdot 2\right)^2 = 1.063$$

The displacement table I-A is multiplied by $S_{IA} = 0.9539''$; one division of the table = 44° ÷ 180 = 14′ 40″ of the arc; 3 divisions = 14′ 40″ × 3 = 44′ of the arc.

The displacement table II-D is multiplied by $S_{IID} = 1.5461''$ and S_{IA} added. One division of the table = 66° ÷ 180 = 22′ of the arc; 2 divisions = 44′ of the arc.

Second stroke:

$$P = \frac{3}{2}; \quad \alpha_{IIA} = 66°; \quad \alpha_{ID} = 44°$$

Strokes:

$$S_{IIA} = 1.5461''; \quad S_{ID} = 0.9539''$$

Coefficients:

$$C_{II} = 3.846; \quad C_I = 6.592; \quad C_V = 1.7596 \frac{1 + P}{1 + 1.0805P} = 1.7$$

The displacement table II-A is multiplied by $S_{IIA} = 1.5461''$, one division of the table = 66° ÷ 180 = 22′ of the arc; 2 divisions = 44′. The displacement table I-D is multiplied by $C_{ID} = 0.9539''$ and S_{IIA} added; one division of the table = 44° ÷ 180 = 14′ 40″ of the arc; 3 divisions = 44′.

The displacement table I-D is multiplied by $C_{ID} = 0.9539''$ and S_{IIA} added; one division of the table = $44° \div 180 = 14'\,40''$ of the arc; 3 divisions = $44'$.

Remarks: the working periods of both strokes are the same, therefore, the deceleration at the first stroke is the same as the acceleration of the second stroke; so there is no variation of the dynamic force at this point.

NOTE: the value of the factor P remains the same for both strokes, because in both cases, the Shape II should have a longer time than the Shape I—then the dynamic forces at the point A (Fig. IV-4) will be of the same magnitude.

EXAMPLE V

Suppose: (a) the movement is of the Type III—without a dwell between two strokes—so the acceleration diagram III should be used; (b) the designer cannot arrange to have the same working period for both strokes.

It means one stroke has a longer working period, which may be called a "slow" stroke and the other has shorter working period, so it may be called a "fast" stroke.

If the same factor P is used for both strokes, the deceleration at the end of the first stroke will not be equal to the acceleration of the second stroke; at that moment, there will be an instantaneous change in magnitude of the acceleration and, therefore, an instantaneous change in magnitude of the dynamic force. If this change is sufficiently large, there is a possibility of noticeable increase of vibrational stresses after that moment. Evidently, different factors P should be used for the first and the second strokes— to eliminate or, at least, to decrease the size of the instantaneous change of the acceleration.

In a previous chapter, it was explained that in case of movements of the Type III, the coefficients of the maximum acceleration depend on the shape of the deceleration diagram and it does not matter if this shape is used for the period of the acceleration or for the period of the deceleration. Also, the equations for C_I and C_{II} are given, as function of the factor P. The first problem is to equalize the maximum accelerations of the Shape II diagram, which are affected by the factor P and by the working periods $\alpha°$, which are different in two strokes; to distinguish them, introduce subscripts for them:

a subscript s — for the slow stroke with a longer working period, and

a subscript f — for the fast stroke with a shorter working period;

it does not matter if the Shape II diagram belongs to the first stroke or to the second.

The problem will be studied for the Modified Sine Diagram III and the results can be applied to other acceleration diagrams easily, because the difference is only in some numerical coefficient in the equations.

For the slow stroke

$$C_{IIS} = 4.8376 \frac{(1 + P_S)^2}{2P_S(1 + 1.0805P_S)} = 2.4188 \frac{(1 + P_S)^2}{P_S(1 + 1.0805P_S)}$$

For the fast stroke

$$C_{IIf} = 2.4188 \frac{(1 + P_f)^2}{P_f(1 + 1.0805P_f)}$$

The general form for the acceleration is

$$C_{II} \cdot S\left(\frac{360}{\alpha}N\right)^2.$$

Write expressions for the acceleration for both strokes and equalize them. The stroke S, the speed N and numerical coefficients can be canceled and the basic equation will be

$$\frac{(1 + P_f)^2}{P_f(1 + 1.0805P_f)\alpha_f^2} = \frac{(1 + P_S)^2}{P_S(1 + 1.0805P_S)\alpha_S^2} \qquad \text{(a)}$$

Reverse the numerators and the denominators

$$\frac{P_f + 1.0805P_f^2}{1 + 2P_f + P_f^2} = \left[\left(\frac{\alpha_S}{\alpha_f}\right)^2 \cdot \frac{P_S(1 + 1.0805P_S)}{(1 + P_S)^2}\right] = K \qquad \text{(b)}$$

After the transformation

$$P_f + 1.0805P_f^2 = K + 2KP_f + KP_f^2$$

$$(1.0805 - K)P_f^2 - (2K - 1)P_f - K = 0$$

$$P_f = \frac{K - 0.5 + \sqrt{0.0805K + 0.25}}{1.0805 - K} \qquad \text{(A)}$$

P_f should be positive, therefore + sign is used before the radical.

If the ratio $\frac{\alpha_S}{\alpha_f}$ is known and the factor P_S is selected, then the factor P_f can be determined from the equation (A); this combination of P_S and P_f means that maximum acceleration of the Shapes II are equal for both strokes for a given ratio $\frac{\alpha_S}{\alpha_f}$.

Then, it is necessary to check what effect this solution will have on maximum acceleration of the Shapes I; it may increase it for the fast stroke to unacceptable magnitude, because it was higher than the acceleration of the Shape II when $P_f = 1$, so α depression of the maximum acceleration of the Shape II will increase it still more.

It is evident that the best possible solution for P_s and P_f could be found only after many calculations based on original selection of the factor P_s; these calculations will be almost useless when a next similar problem comes up.

A better practical solution is to prepare charts which would permit a quick appraisal of the selected factor P_s on the value of the maximum acceleration of the Shape I diagram, especially for the fast stroke. Calculations can be made for different ratios of the working periods $\frac{\alpha_s}{\alpha_f}$, which can be used as parameters on the chart.

One group of curves will show relations between P_s and P_f, which produce equal maximum acceleration of the diagram of the Shape II for both strokes for certain values of the ratio $\frac{\alpha_s}{\alpha_f}$.

The other group of curves may show comparative values of maximum acceleration for diagram of the Shape I.

These two groups of curves can have common ordinates, but with different items on the base or abscissa line; in that way they will make one chart. The process of the preparation of the necessary data is offered for the Modified Sine Diagram III.

It is possible to establish quite narrow limits for the factor P_s: the slow stroke with a longer working period α_s will have smaller maximum + and − acceleration than the fast stroke. To equalize the accelerations of both strokes, it is necessary to increase the maximum acceleration of the Shape II of the slow stroke; it means P_s should be less than unity, so the calculations are made for $P_s = 0.3 \sim 1.0$.

As it was mentioned before, the factor P_f can be found from the equation (A); the item K, entering into it, can be presented as

$$K = \left(\frac{\alpha_s}{\alpha_f}\right)^2 \cdot M_s \text{ where } M_s = \frac{P_s(1 + 1.0805P_s)}{(1 + P_s)^3} \qquad (c)$$

M_s can be introduced into expressions for coefficients C_I and C_{II} for the slow stroke, because it is calculated for it.

$$C_{IIS} = \frac{2.4188}{M_s} \qquad (d)$$

$$C_{IS} = \frac{2.764P_s}{M_s} \qquad (e)$$

For the fast stroke, M_f is different because P_s is replaced by P_f; when it is introduced into expressions for coefficients C_I and C_{II} in this case, these expressions have the form

$$C_{IIf} = \frac{2.4188}{M_f} \tag{f}$$

$$C_{If} = \frac{2.764P_f}{M_f} \tag{g}$$

The coefficients C_{IIS} and C_{IIf} [equations (d) and (f)] are different in form, but they produce the maximum accelerations of the same magnitude, because the factor P_f, different than P_s, is determined from the equations (a) and (A), which are based on the equality of maximum accelerations produced by diagrams of the Shape II.

The coefficients C_{IS} and C_{If} [equations (e) and (g)] are also different, but the maximum accelerations produced by diagrams of the Shape I are affected by an additional cause—namely, by the difference in working periods α_s and α_f.

The maximum acceleration during the slow stroke is proportional $C_{IS} \cdot \dfrac{1}{\alpha_s{}^2}$ and the maximum acceleration during the fast stroke is proportional to $C_{If} \cdot \dfrac{1}{\alpha_f{}^2}$.

The ratio $\dfrac{\text{Max. acceleration for the fast stroke}}{\text{Max. acceleration for the slow stroke}} = \dfrac{C_{If}}{C_{IS}} \cdot \left(\dfrac{\alpha_s}{\alpha_f}\right)^2$

indicates in dimensionless way the difference between maximum accelerations. To express this difference in a way suitable for plotting on the chart, this ratio should be multiplied by C_{IS}.

Then the expression $C_{If} \cdot \left(\dfrac{\alpha_s}{\alpha_f}\right)^2$ will be like an artificial coefficient of maximum acceleration for the faster stroke, which can be compared with the coefficient C_{IS} for the slow stroke and plotted on the chart on the same scale as C_{IS}.

The Chart 22 is for Modified Sine and the Chart 23 for Modified Trapezoid Diagram. As it was said before, the ratios $\dfrac{\alpha_s}{\alpha_f}$ are used as parameters and they are marked on the charts.

Dotted lines on the right side of the charts show coefficients C_I and C_{II} for the slow stroke; they represent a base for comparison of maximum accelerations for the slow stroke with maximum accelerations of the fast stroke. These charts permit a quick appraisal of comparative magnitude of accelerations for both strokes.

To illustrate the use of the charts, suppose the ratio of working periods $\frac{\alpha_s}{\alpha_f} = 1.2$. Tentatively, select the factor P_s for the slow stroke.

On the left side of the Chart 22 draw a vertical line from $P_s = 0.7$ to the intersection with the P_f curve for $\frac{\alpha_s}{\alpha_f} = 1.2$; P_f is approximately equal 1.415.

From the intersection, draw a horizontal line to the corresponding curve on the right side of the chart and a vertical line to the scale on the right side. Thus, a maximum acceleration of the fast stroke is presented by the number 9.17. On the ordinates on the right side of the chart representing P_s, draw a horizontal line from the selected $P_s = 0.7$; it intersects the dotted lines C_{IS} and C_{IIS}; the intersections permit to find $C_{IIS} = 5.65$ and $C_{IS} = 4.55$.

These figures can be tabulated as giving comparative values of accelerations:

Slow stroke—max. acceleration Type I is represented by $C_{IS} = 4.55$

max. acceleration Type II is represented by $C_{IIS} = 5.65$,

Fast stroke—max. acceleration Type I is represented by $C_{If} \cdot \left(\frac{\alpha_s}{\alpha_f}\right)^2 =$

$= 9.17$

If the relative values represented by these figures are not satisfactory, another P_s can be tried. To decrease the maximum acceleration during the fast stroke—the factor P_s should be smaller; at the same time the maximum acceleration of the Shape II will increase and be closer to the $C_{IIS} = C_{IIf}$. The decision depends on the judgment of the designer.

The charts are made for maximum $\frac{\alpha_s}{\alpha_f} = 1.3$; even at this ratio, the balancing of acceleration is not practical.

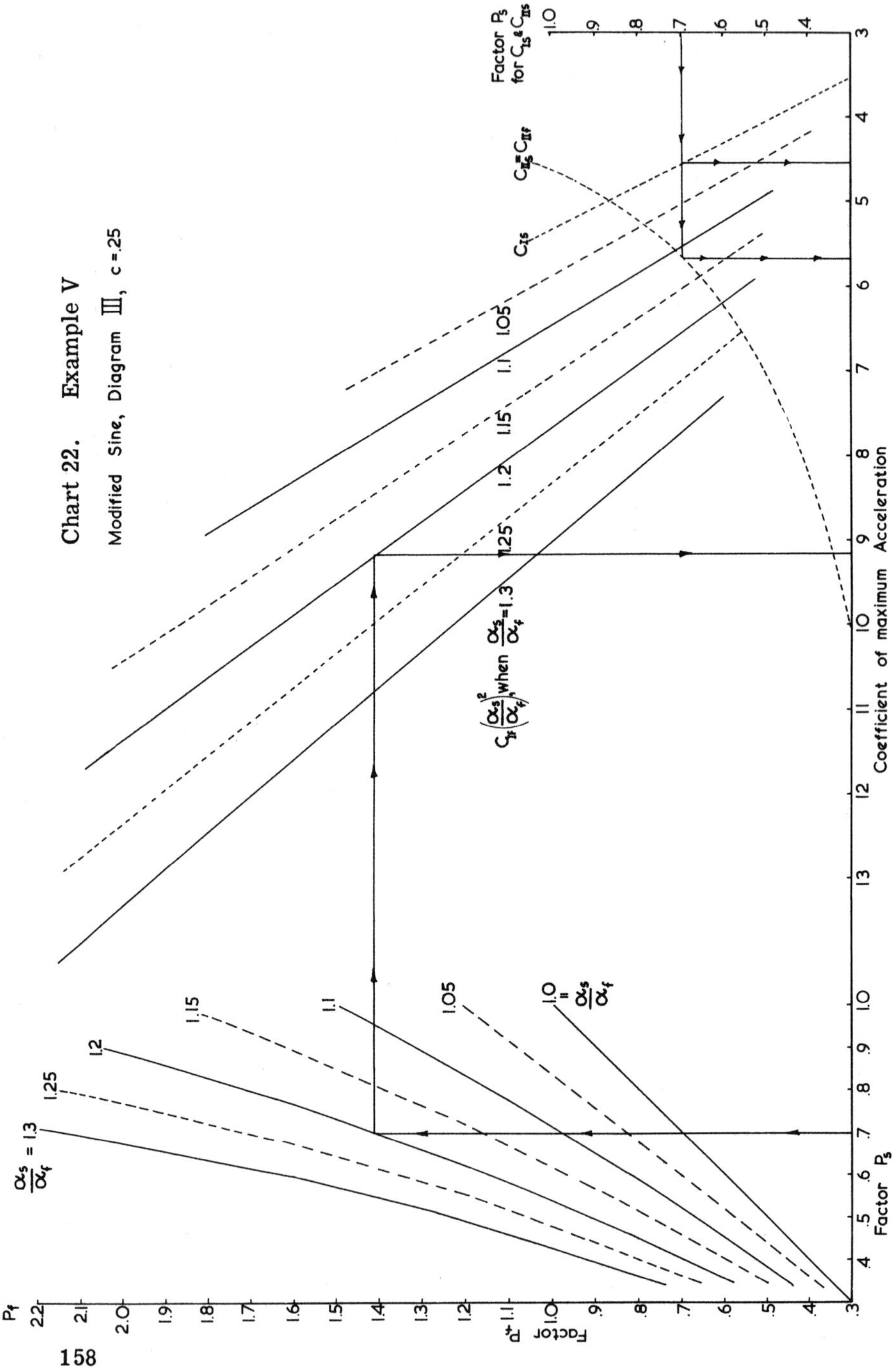

Chart 22. Example V

Modified Sine, Diagram III, c = .25

Chart 23. Example V

Modified Trapezoid, Diagram III, b = .25.

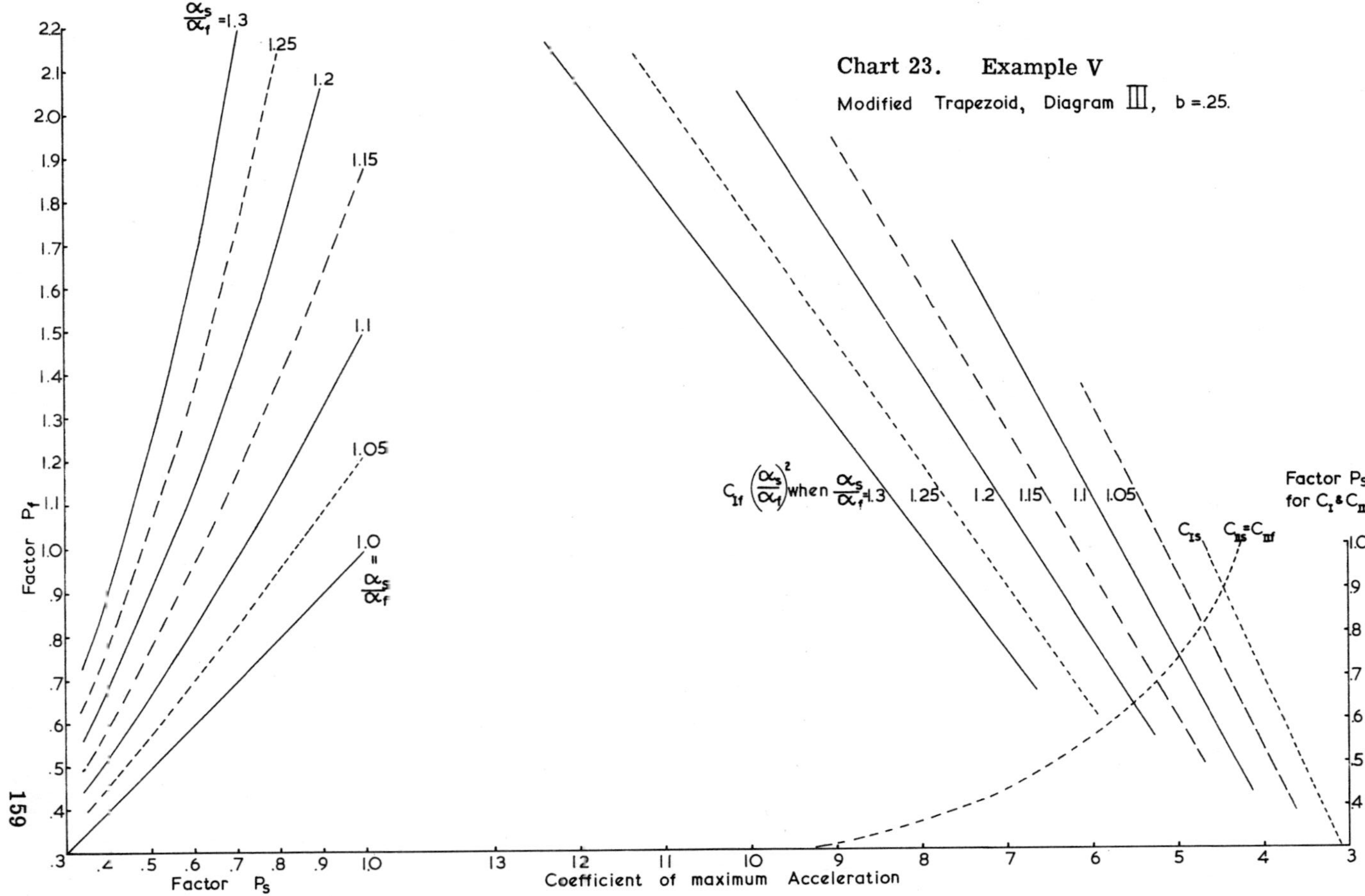

EXAMPLE VI

Plate cam with a spring holding the cam follower against the cam surface.

The work stroke is produced by the spring when the cam follower moves toward the center of the cam (a drop)—to prevent a serious jam, if a part pushed by the cam follower system cannot enter the holder.

The work stroke is made in two steps:

(1) the first step starts from a dwell; the stroke $S_1 = 0.625''$ and the working period $\alpha_1^\circ = 60^\circ$; it ends in a 40° dwell;

(2) the second step starts after this 40° dwell and it is followed by a return stroke with a dwell between them; the stroke $S_2 = 1.75''$; 210° are allotted for this second step and for the return stroke.

The return stroke is $S = S_1 + S_2 = 2.375''$ and it ends in a dwell $= 50^\circ$.

Select a Modified Trapezoid acceleration. For the first step of the work stroke, the Diagram I is used, as it is the movement of the Type I—between two dwells. For the second step and for the return stroke, the Diagram III should be used.

The maximum acceleration of the second step (at the end of it) should be equal or close to the maximum acceleration of the return stroke. To explain the problem, the diagrams of displacements and of the acceleration are drawn.

Markings of the acceleration diagrams with letters A and D and also IA, ID and IIA, IID helps to select proper displacement table. Markings "drop" and "rise" indicate if the displacement table starts from the maximum radius or from the minimum radius of the cam.

Work Stroke—a drop.

First step: stroke $S_1 = 0.625''$; working period $\alpha_1 = 60^\circ$; the factor $P_1 = 2/3$ is selected—to decrease the acceleration of the cam follower, because the spring has to provide the acceleration.

Select Modified Trapezoid Diagram I.

The displacement during the acceleration period

$$S_A = \frac{S_1}{1 + P} = \frac{3}{5} S_1 = \frac{3}{8}''$$

It is used to multiply the data from the displacement table I-A.

The displacement during the deceleration period

$$S_D = S_1 - S_A = 1/4''$$

Multiply the data from the table I-D by S_D and add $3/8''$ to each product.

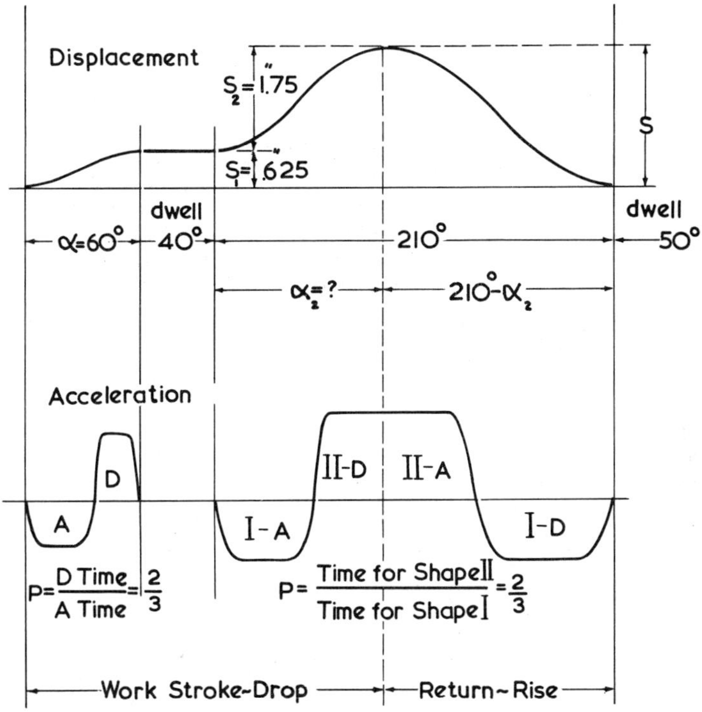

Fig. IV - 5

Working periods: for acceleration

$$\alpha_A = \frac{\alpha_1}{1 + P} = \frac{3}{5} \cdot 60° = 36°;$$

one division in the table $= 36° \div 180 = 0.2° = 12'$ of the arc.

— for deceleration $\alpha_D = 60 - \alpha_A = 24°$; one division in the table $= 24° \div 180 = 1440' \div 180 = 8'$ of the arc.

Maximum acceleration coefficients:

$$C_A = 4.888 \frac{1 + P}{2} = 4.888 \cdot \frac{5}{6} = 4.075;$$

$$C_D = 4.888 \frac{1 + P}{2P} = 4.888 \cdot \frac{5}{4} = 6.11.$$

Second step: stroke $S_2 = 1.75''$; the factor

$$P = \frac{\text{Time for Shape II}}{\text{Time for Shape I}} = \frac{2}{3}$$

This step is combined with the return stroke $S = 2.375''$; there is no dwell between them and two of them have the total working period $\alpha_2^\circ + \alpha_3^\circ = 210°$.

Select Modified Trapezoid Diagram III; the factor $P = 2/3$ is used for both strokes — to decrease the required spring tension. The maximum deceleration at the end of the second step of the work stroke should be equal to the maximum acceleration at the beginning of the return stroke.

Both adjoining periods have the acceleration diagram of the Shape II, therefore, their coefficients C_{II} are alike.

Then

$$C_{II} \cdot S_2 \left(\frac{360}{\alpha_2}\right)^2 N^2 = C_{II} \cdot S \left(\frac{360}{\alpha_3}\right)^2 N^2;$$

$$\frac{S_2}{\alpha_2^2} = \frac{S}{\alpha_3^2}; \quad \left(\frac{\alpha_3}{\alpha_2}\right)^2 = \frac{S}{S_2} = \frac{2.375''}{1.75''} = 1.3(571428)\ldots.$$

$$\frac{\alpha_3}{\alpha_2} = \sqrt{\frac{S}{S_2}} = 1.64969 = 1.165; \qquad \alpha_3 = 1.165\alpha_2$$

The sum

$$\alpha_2 + \alpha_3 = 210°;$$

$$2.165\alpha_2 = 210°;$$

$$\alpha_2 = \frac{210°}{2.165} = 96.999° = 97°$$

$$\alpha_3 = 210° - 97° = 113°$$

The second step of the work stroke and the return stroke have the same factor $P = 2/3$; therefore, their coefficients C_I and C_{II} are alike.

$$C_I = 4.89\,\frac{1 + P}{2} \cdot \frac{1 + P}{1 + 1.0869P} = 3.94;$$

$$C_{II} = 4.39\,\frac{1 + P}{2P} \cdot \frac{1 + P}{1 + 1.0869P} = 5.30.$$

EXAMPLE VII

A rotary movement should be produced with the variation of the velocity during one cycle; the requirements are:

(1) 24% of a stroke S should be produced during 80% of the cycle time T at a constant velocity.

(2) 76% of the stroke S should be made during 20% of the time T with a varying velocity and with a gradually changing acceleration; at the end of this period, the next cycle starts with the constant velocity described in § 1.

The movement is made by a barrel cam which rotates on 360° during one indexing cycle with the total stroke S of the driven part. The stroke S is measured in degrees $= \frac{360°}{n}$, where n is an integer as 4, 5, 6, etc., depending on the design of the driven part.

Actually, the desired movement is similar to usual indexing by an intermittent motion, when a dwell period is replaced by a period of constant velocity.

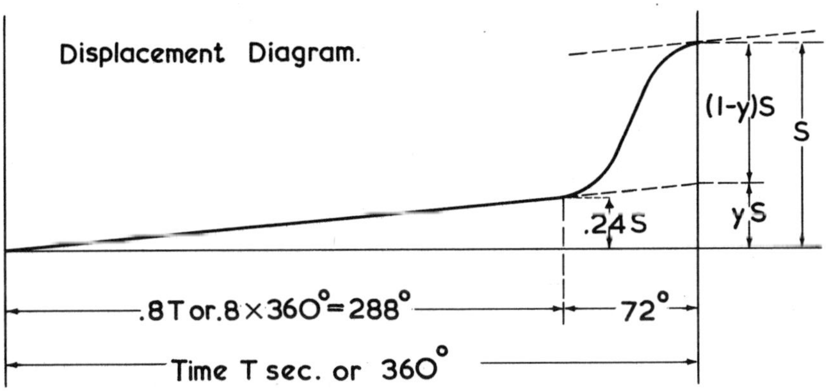

Fig. IV - 6

Increments of the displacement created by the constant velocity continue during the whole cycle, so the total displacement due to the constant velocity should be determined.

$$\frac{YS}{T} = \frac{0.24S}{0.8T}; \qquad Y = 0.3.$$

Then the displacement to be produced during the period of the varying velocity is $(1 - Y)S = 0.7S$.

Modified Trapezoid Diagram I is selected with the factor $P = 1$, so during the acceleration and during the deceleration periods $S_A = S_D = 0.35\ S$ and $\alpha_A = \alpha_D = \dfrac{72°}{2} = 36°$ on the cam.

Each division of the displacement table $= 36° \div 180 = 0.2° = 12'$.

The Table I-A is multiplied by $0.35S$; the Table I-D is also multiplied by $0.35S$ and to each product S_A is added. So, the displacements are tabulated for the period of 72°.

During the same period, the uniform velocity produces $(Y - 0.24)S = 0.06S$ and for each division of the displacement table, it adds an increment $0.06 \div 360 = 0.0001(6)\ldots S = \dfrac{S}{6000}$. This item should be added to every division of the table prepared for the period of the varying velocity. The period of the constant velocity can be cut in the cam as a constant slope of

$$0.24S \div 288° = \frac{S}{1200} \text{ per degree of the cam rotation.}$$

The acceleration is affected only by the part of the stroke S, namely by $(1 - Y)S = 0.7\ .\ S$

Max. angular acceleration

$$\Omega = 4.89\ .\ 0.7S \cdot \left(\frac{360}{72°}\ .\ N\right)^2 \text{ rad./sec.}^2$$

S should be expressed in radians, as $\dfrac{2\pi}{n}$

EXAMPLE VIII

The driven part is indexed on 36° per cycle or $S = \dfrac{360°}{n = 10} = \dfrac{2\pi}{10}$ radians. A barrel cam is used; it makes one revolution per cycle and time per cycle is given 2.2 sec. The dwell occupies 1.4 sec., therefore, the indexing time is 0.8 sec. During the first half of indexing time, or during 0.4 sec., the displacement of the driven part is $\frac{2}{3}$ (or 24° or $\dfrac{2\pi}{15}$ radians). After that period, a smallest possible velocity is maintained as long as possible and then it is brought down to zero.

So, the time allotted to the constant velocity and to the final deceleration is 0.4 sec. It is desirable to have the maximum acceleration equal to the maximum deceleration. Modified Sine Diagram I is selected because it has lower coefficient of maximum velocity, $C_V = 1.76$, while Modified

Trapezoid and Sine with C = 0.5 have $C_V = 2$. During calculations, retain C_V instead of its value, to be certain that the selection of the lower C_V is really advantageous.

Due to the complexity of the problem, diagrams are drawn for the displacement, the velocity and for the acceleration.

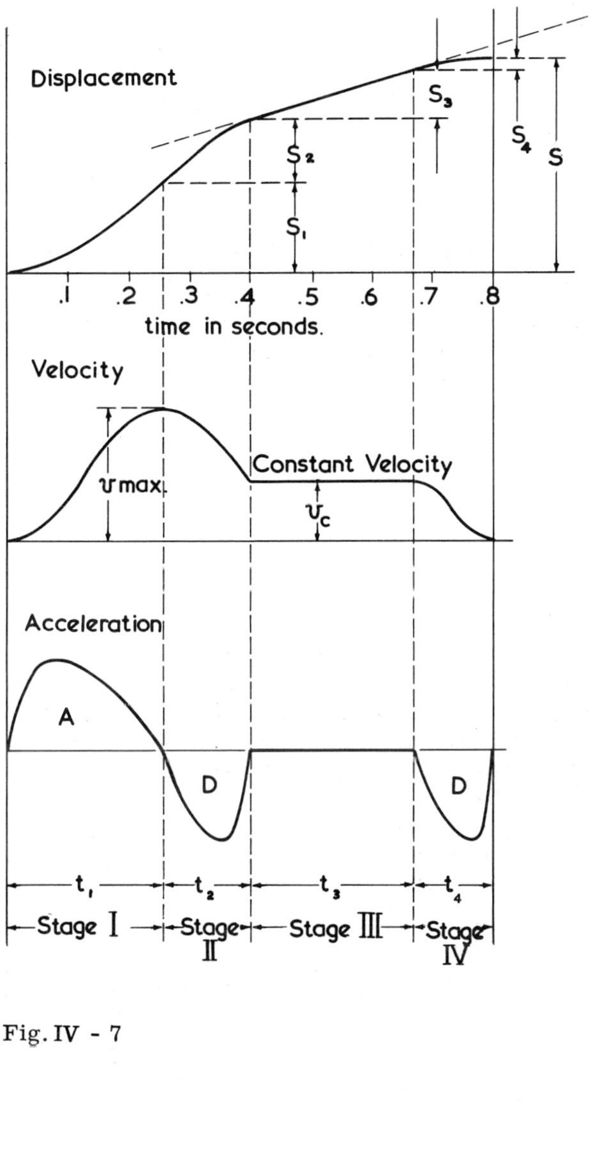

Fig. IV - 7

Represent the requirements:

Stages I and II:

$t_1 + t_2 = 0.4$ sec.; or $t_2 = 0.4 - t_1$ (a)

$S_1 + S_2 = 24°$; or $S_2 = 24 - S_1$ (b)

max. v_1 = max. v_2. (c)

Stages III and IV:

$t_3 + t_4 = 0.4$ sec. or $t_4 = 0.4 - t_3$ (d)

$S_3 + S_4 = 12°$ or $S_4 = 12 - S_3$ (e)

max. $v_4 = v_C$ (constant velocity) (f)

Maximum acceleration and maximum deceleration in Stages I, II and IV should be equal (g)

Due to the presence of many unknown items, the equations should have complicated forms, therefore, known items are presented by their actual values as the total time of one cycle = 0.8 sec. and the total stroke = 36°. The maximum velocity coefficient will be retained in equations until they become too complicated. The solution of the problem is based on requirements of equal velocities requirements (c) and (f) and on requirements expressed in (g).

Equalize velocities

Stages I and II

The maximum velocity at the end of the Stage I is $v_1 = C_V \cdot \frac{S_1}{t_1}$ (h)

The Stage II is similar to the case described in the Example VII; its maximum velocity is affected by the constant velocity during the Stage III and by the change of the velocity, due to the deceleration. To find the velocity change during the deceleration, it is necessary to determine the displacement during the Stage II.

Designate the component of S_2 caused by the deceleration as S_{2D}; the displacement due to the presence of the constant velocity during the Stage III is $v_C \cdot t_2$. Then

$$S_2 = S_{2D} + v_C \cdot t_2 \quad \text{and} \quad S_{2D} = S_2 - v_C \cdot t_2$$

Using the equations (a) and (b), express S_{2D} in terms of S_1 and t_1

$$S_{2D} = 24 - S_1 - v_C \cdot (0.4 - t_1) \qquad \text{(i)}$$

Then the change of velocity during the Stage II is $C_V \cdot \frac{S_{2D}}{t_2}$, so the velocity at the beginning of the Stage II is

$$v_{II} = v_C + C_V \cdot \frac{S_{2D}}{t_1} = v_C + C_V \frac{24 - S_1 - v_C(0.4 - t_1)}{0.4 - t_1} =$$

$$= v_C + C_V \frac{24 - S_1}{0.4 - t_1} - C_V \cdot v_C =$$

$$= C_V \cdot \frac{24 - S_1}{0.4 - t_1} - (C_V - 1) \cdot v_C \qquad \text{(j)}$$

To meet the requirement expressed in the equation (c), equalize the equations (h) and (j).

$$C_V \cdot \frac{S_1}{t_1} = C_V \frac{24 - S_1}{0.4 - t_1} - (C_V - 1)v_C;$$

$$\frac{S_1}{t_1} = \frac{24 - S_1}{0.4 - t_1} - \frac{C_V - 1}{C_V} \cdot v_C \qquad (1)$$

The maximum velocity coefficient C_V is the same for both stages, because the same acceleration diagram is used.

Stages III and IV.

The velocity at the beginning of the Stage IV is v_C and it is brought to zero at the end, due to deceleration; this change of velocity is $v_{IV} = C_V \cdot \frac{S_4}{t_4}$.

Then, to satisfy equation (f)

$$v_C = C_V \cdot \frac{S_4}{t_4} = C_V \cdot \frac{12 - S_3}{0.4 - t_3} \qquad \text{(k)}$$

The displacement during the Stage III is $S_3 = v_C \cdot t_3$ (l)

$$v_C = C_V \cdot \frac{12 - v_C t_3}{0.4 - t_3};$$

$$v_C(0.4 - t_3) = C_V(12 - v_C t_3);$$

$$v_C(0.4 - t_3 + C_V \cdot t_3) = C_V \cdot 12;$$

$$v_C = \frac{C_V \cdot 12}{0.4 + (C_V - 1) \cdot t_3} \qquad (2)$$

<u>Equalize accelerations</u>

Max. acceleration during the Stage I is

$$A_1 = C_A \cdot \frac{S_1}{t_1{}^2} \tag{m}$$

Max. deceleration during the Stage II is

$$D_{II} = C_A \cdot \frac{S_{2D}}{t_2^2} = C_A \cdot \frac{24 - S_1 - v_C(0.4 - t_1)}{(0.4 - t_1)^2} =$$

$$= C_A \left[\frac{24 - S_1}{(0.4 - t_1)^2} - \frac{v_C}{0.4 - t_1} \right] \tag{n}$$

Max. deceleration during the Stage IV is

$$D_{IV} = C_A \cdot \frac{S_4}{t_4^2} = C_A \cdot \frac{12 - S_3}{(0.4 - t_3)^2}.$$

From the equation (k)

$$\frac{12 - S_3}{0.4 - t_3} = \frac{v_C}{C_V},$$

therefore

$$D_{IV} = C_A \cdot \frac{v_C}{C_V(0.4 - t_3)} \tag{o}$$

Equations (m), (n), and (o) should be equal, therefore,

$$\frac{S_1}{t_1^2} = \frac{24 - S_1}{(0.4 - t_1)^2} - \frac{v_C}{0.4 - t_1} \tag{3}$$

$$\frac{S_1}{t_1^2} = \frac{v_C}{C_V(0.4 - t_3)} \tag{4}$$

There are four equations (1), (2), (3), and (4) with four unknown items: S_1, t_1, v_C and t_3. Therefore, they can be found.

Then the equations (a), (b), (d), (e) and (1) permit to find the remaining five unknown items: S_2, S_3, S_4, t_2 and t_1.

Multiply the equation (3) by $(0.4 - t_1)$

$$\frac{S_1(0.4 - t_1)}{t_1^2} = \frac{24 - S_1}{0.4 - t_1} - v_C$$

Subtract from equation (1)

$$\frac{S_1}{t_1} - \frac{S_1(0.4 - t_1)}{t_1^2} = \frac{24 - S_1}{0.4 - t_1} - \frac{C_V - 1}{C_V} \cdot v_C - \frac{24 - S_1}{0.4 - t_1} + v_C =$$

$$= v_C\left(1 - \frac{C_V - 1}{C_V}\right) = \frac{v_C}{C_V};$$

$$S_1 \cdot \frac{2t_1 - 0.4}{t_1^2} = \frac{v_C}{C_V};$$

$$S_1 = v_C \frac{t_1^2}{C_V(2t - 0.4)} \qquad (5)$$

Introduce S_1 from the equation (5) into the equation (4)

$$\frac{v_C t_1^2}{C_V(2t_1 - 0.4)t_1^2} = \frac{v_C}{C_V(0.4 - t_3)};$$

$$2t_1 - 0.4 = 0.4 - t_3$$

$$t_3 = 0.8 - 2t_1 \qquad (6)$$

Introduce t_3 from the equation (6) into the equation (2)

$$v_C = \frac{12C_V}{0.8C_V - 4 - 2(C_V - 1)t_1}; \qquad \frac{C_V}{v_C} = \frac{0.8C_V - 4 - 2(C_V - 1)t_1}{12} \qquad (7)$$

Introduce S_1 from the equation (5) into the equation (1)

$$v_C \frac{t_1^2}{C_V(2t_1 - 0.4)t_1} = \frac{24 - v_C \dfrac{t_1^2}{C_V(2t_1 - 0.4)}}{0.4 - t_1} - v_C \frac{C_V - 1}{C_V}$$

Multiply both sides of this equation by

$$\frac{C_V}{v_C} \cdot (2t_1 - 0.4) \cdot (0.4 - t_1)$$

after transformations

$$2(C_V - 1)t_1^2 - \left[1.2(C_V - 1) - 48\frac{C_V}{v_C} + 0.4\right].\ t_1 -$$

$$\left[-\frac{C_V}{v_C}9.6 - 0.16(C_V - 1)\right] = 0 \tag{8}$$

If $\frac{C_V}{v_C}$ from the equation (7) is introduced, this equation will have only one unknown—t_1—so it can be found.

$$t_1^2 - \left(0.4 + 0.2\frac{C_V}{C_V - 1}\right)t_1 + \frac{1}{3}\left(0.08 + 0.16\frac{C_V}{C_V - 1}\right) = 0 \tag{8$_1$}$$

$$t_1 = 0.2 + 0.1\frac{C_V}{C_V - 1} \pm \sqrt{\left(\frac{0.1C_V}{C_V - 1}\right)^2 - \frac{0.01(3)....}{C_V - 1}} \tag{8$_2$}$$

On Fig. IV-7, it can be seen that t_1 is smaller than 0.4 sec., so — sign should be used.

To check—try two acceleration diagrams

Modified Trapezoid; $C_V = 2$;

$$t_1 = 0.4 - \sqrt{0.02(6)...} = 0.4 - 0.1633 = 0.2367 \text{ sec.}$$

Modified Sine; $C_V = 1.7596$;

$$t_1 = 0.43165 - \sqrt{0.0361079} = 0.43165 - 0.19002 = 0.2416 \text{ sec.}$$

From the equation (7) the constant velocity can be found, but the expression is very complicated unless C_V is selected.

From the equation (6) t_3 can be found and S_1 — from the equation (5); again the expressions are complicated until a definite C_V is introduced.

Determine v_C from the equation (7)

Mod. Sine $v_C = 32.9616°/\text{sec.}$

Mod. Trapezoid $v_C = 33.03055°/\text{sec.}$

the difference is only 0.21°/sec.

From the equation (6) $t_3 = 0.8 - 2t_1$

Mod. Sine $t_3 = 0.31674$ sec.

Mod. Trapezoid $t_3 = 0.3266$ sec.

From the equation (a) $t_2 = 0.4 - t_1$

Mod. Sine $t_2 = 0.15837$ sec.

Mod. Trapezoid $t_2 = 0.1633$ sec.

From the equation (d) $t_4 = 0.4 - t_3$

Mod. Sine $t_4 = 0.08326$ sec.

Mod. Trapezoid $t_4 = 0.734$ sec.

Determine S_1 from the equation (5)

Mod. Sine $S_1 = 13.1357°$

Mod. Trapezoid $S_1 = 12.6063°$

Then, $S_2 = 24° - S_1$;

Mod. Sine $S_2 = 10.8643°$

Mod. Trapezoid $S_2 = 11.3937°$

Determine S_3 from the equation (1) — $S_3 = v_C \cdot t_3$

Mod. Sine $S_3 = 32.9616 \times 0.31674 = 10.4403°$

Mod. Trapezoid $S_3 = 33.03055 \times 0.3266 = 10.7878°$

Then $S_4 = 12° - S_3$;

Mod. Sine $S_4 = 1.5597°$

Mod. Trapezoid $S_4 = 1.2122°$

From the equation (j) determine S_{2D}—for the calculation of the deceleration during the Stage II;

Mod. Sine $S_{2D} = 5.6442°$

Mod. Trapezoid $S_{2D} = 5.9999°$

Check velocities caused by the acceleration.

Stage I $v_1 = C_V \cdot \dfrac{S_1}{t_1}$ Stage II $v_2 = C_V \dfrac{S_{2D}}{t_2} + v_C$;

Mod. Sine $v_1 = 95.66°/\text{sec.}$ $v_2 = 95.66°/\text{sec.}$

Mod. Trapezoid $v_1 = 106.51°/\text{sec.}$ $v_2 = 106.51°/\text{sec.}$

Check acceleration $\div C_A$

Stage I $\dfrac{S_1}{t_1^2}$;

Mod. Sine $225°/\text{sec.}^2$

Mod. Trapezoid $225°/\text{sec.}^2$

Stage II $\dfrac{S_{2D}}{t_2^2}$

Mod. Sine $225°/\text{sec.}^2$

Mod. Trapezoid $225°/\text{sec.}^2$

Stage IV $\dfrac{S_4}{t_4^2}$

Mod. Sine $225°/\text{sec.}^2$

Mod. Trapezoid $225°/\text{sec.}^2$

Actual acceleration $= C_A \dfrac{S}{t}$

Mod. Sine $5.528 \times 225 = 1243.8°/\text{sec.}^2 = 21.7\ \text{rad./sec.}^2$

Mod. Trapezoid $4.888 \times 225 = 1099.8°/\text{sec.}^2 = 19.2\ \text{rad./sec.}^2$

Comparison of the Sine and Trapezoid diagram

	Mod. Sine	Mod. Trapezoid
	Seconds	Seconds
t_1	0.24163	0.2367
t_2	0.1584	0.1633
t_3	0.3167	0.3266
t_4	0.0833	0.0734
	Degrees	Degrees
S_1	13.1357	12.6063
S_{2D}	5.6442	5.9999
S_2	10.8643	11.3937
S_3	10.4403	10.7878
S_4	1.5597	1.2122
v_C	32.9616	33.0306°/sec.
Accel	21.7	19.2 rad./sec.^{2}

The difference between two diagrams is insignificant:

Mod. Trapezoid has a longer time for the uniform velocity and lower max. acceleration.

Mod. Sine has a lower constant velocity.

This example demonstrates the usefulness of the coefficients C_V and C_A and the convenience of the writing of equations in the form described on pp. 11 & 12 of the Chapter I.

The displacement tables I-A and I-D are used and multiplied by S_1 and S_4. During the Stage II, the I-D table is multiplied by S_{2D} and increments caused by the constant velocity added.

During the Stage III, the increments of displacements = S_3 divided by $\frac{t_3}{0.4} \cdot 12°$; the result is in degrees of the driven part movements by degree of the cam rotation.

The conversion of S_1, S_2, S_3, and S_4 from degrees to lineal displacement on the cam depends on physical dimensions of the drive.

EXAMPLE IX

Requirements:

(1) a bed of a printing press has a reciprocating motion;

(2) the printing cylinder rotates uniformly;

(3) one cycle of bed motion is produced while the cylinder makes one revolution;

(4) during the forward stroke, the bed has a period of uniform velocity, which is equal to the velocity on the periphery of the printing cylinder;

(5) the return stroke may be made during a shorter time;

(6) the deceleration at the end of either stroke should be equal or close to the acceleration of the next stroke;

(7) the bed of the press completes one cycle while a cam driving it rotates on 360°.

Nomenclature:

$\alpha°$ — period of the acceleration and of the deceleration during the forward stroke

$\gamma°$ — period of uniform velocity during the forward stroke

$\beta°$ — period of the return stroke

S — displacement during the varying velocity period of the forward stroke

f — displacement during the uniform velocity period

F = f + S total stroke

v_f — max. velocity during the forward stroke = uniform velocity

v_2 — max. velocity during the return stroke

A_f — max. acceleration or deceleration during the forward stroke

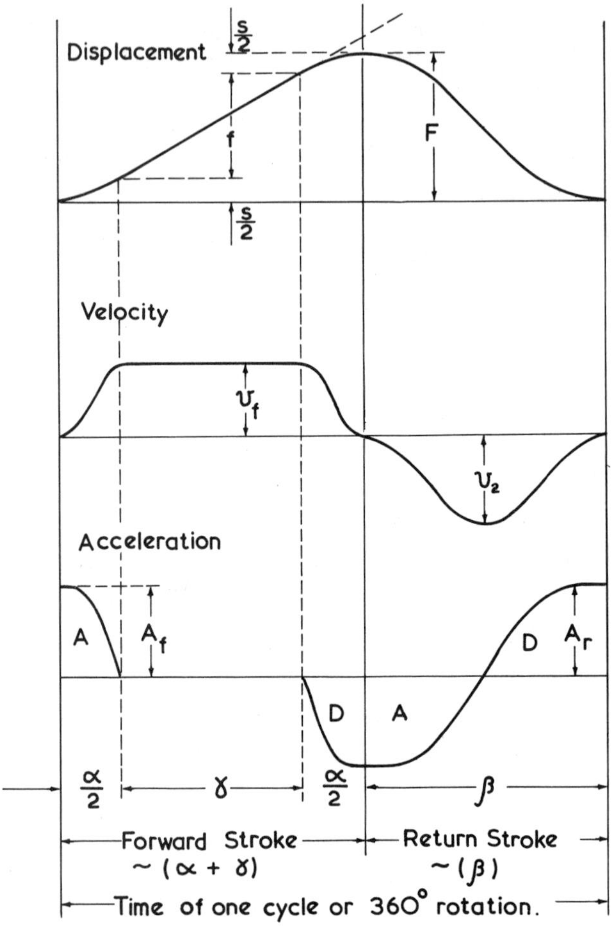

Fig. IV - 8

A_r — the same—during the return stroke

D — the diameter of the printing cylinder

N — speed of the press or the number of cycles per second.

The velocity at the periphery of the printing cylinder is πDN;

$$v_f = \pi DN = C_v \, . \, S \frac{360}{\alpha^\circ} N;$$

Fig. IV - 9. Multicolor Offset Press
(Courtesy of Ferguson Machine Co., Division of UMC Industries)

$$S = \frac{\pi D}{C_V} \cdot \frac{\alpha^\circ}{360} \qquad (1)$$

During the period of the uniform velocity, the following ratios are equal

$$\frac{\gamma^\circ}{360} = \frac{f}{\pi D};$$

therefore,

$$\gamma^\circ = \frac{f}{\pi D} \cdot 360 \qquad (2)$$

$$\alpha^\circ + \beta^\circ = 360 - \gamma = 360\left(1 - \frac{f}{\pi D}\right) = 360\,\frac{\pi D - f}{\pi D}$$

$$\beta^\circ = 360\,\frac{\pi D - f}{\pi D} - \alpha^\circ \qquad (3)$$

The maximum acceleration and the maximum deceleration of both strokes should be equal:

$$A_f = A_r \quad \text{or} \quad C_A S\left(\frac{360}{\alpha}N\right)^2 = C_A(f + S)\left(\frac{360}{\beta}N\right)^2;$$

$$\left(\frac{\beta}{\alpha}\right)^2 = 1 + \frac{f}{S} \qquad (4_1)$$

Introduce S from the equation (1)

$$\beta^2 = \alpha^2 + \frac{f}{\pi D} \cdot C_V \cdot 360 \cdot \alpha \qquad (4_2)$$

Combine equations (3) and (4_2)

$$\alpha^2 + \frac{f}{\pi D} \cdot C_V \cdot 360 \cdot \alpha = 360^2 \cdot \frac{(\pi D - f)^2}{(\pi D)^2} + \alpha^2 - 2\alpha\, 360\, \frac{\pi D - f}{\pi D};$$

$$\alpha = 360\,\frac{(\pi D - f)^2}{\pi D[2\pi D + f(C_V - 2)]} \qquad (5)$$

Introduce α from the equation (5) into the equation (3)

$$\beta = 360\,\frac{\pi D - f}{\pi D} \cdot \frac{\pi D + f(C_V - 1)}{2\pi D + f(C_V - 2)} \qquad (6)$$

So, all angles α, β and γ are expressed in terms of D, f and C_V (in equations (2), (5), and (6)) which are either given or can be selected by the designer. The equations are written for conditions of equality of the maximum acceleration and the deceleration.

There should be some relation between f and D, which is the best. The most desirable aim is to have the lowest acceleration.

The expression for the maximum acceleration can be written for N = 1

$$A = C_A \, . \, S \, . \left(\frac{360}{\alpha}\right)^2.$$

Introduce S from the equation (1)

$$A = \frac{C_A}{C_V} \, . \, \pi D \frac{360}{\alpha}$$

Introduce α from the equation (5)

$$A = \frac{C_A}{C_V} \cdot \left[\frac{C_V \, . \, f \, . \, (\pi D)^2}{(\pi D - f)^2} + \frac{2(\pi D)^2}{\pi D - f}\right].$$

To find the maximum or minimum of A, equalize the first derivative of this expression to zero.

Consider f as a constant and D as a variable

$$\frac{dA}{dD} = \frac{C_A}{C_V} \cdot \left[\frac{(\pi D - f)^2 \, . \, C_V \, . \, f 2\pi^2 D - C_V \, . \, f(\pi D)^2 \, 2(\pi D - f)\pi}{(\pi D - f)^4} + \right.$$

$$\left. + \frac{(\pi D - f)^2 2\pi^2 D 2D - 2(\pi D)^2 \pi}{(\pi D - f)^2}\right] = 0.$$

$\frac{C_A}{C_V}$, $(\pi D - f)$ and $2\pi^2 D$ are not equal zero and $(\pi D - f)$ cannot be infinity; therefore, some of them can be eliminated from this equation.

$$\frac{(\pi D - f)^2 C_V \, . \, f - (\pi D - f)\pi C_V \, . \, fD}{(\pi D - f)^2} + 2(\pi D - f) - \pi D = 0;$$

$$C_V \, . \, f - \frac{\pi D \, . \, C_V \, . \, f}{\pi D - f} + 2(\pi D - f) - \pi D = 0;$$

$$(\pi D)^2 - 3\pi D \, . \, f - (C_V - 2)f^2 = 0;$$

$$\left(\frac{\pi D}{f}\right)^2 - 3\frac{\pi D}{f} - (C_V - 2) = 0;$$

$$\frac{\pi D}{f} = 1.5 \pm \sqrt{2.25 + C_V - 2} = 1.5 \pm \sqrt{C_V + 0.25}$$

When $C_V = 2$; $\frac{\pi D}{f} = 1.5 \pm 1.5 = 3.$ or 0.

evidently + sign should be used.

$$\frac{\pi D}{f} = 1.5 + \sqrt{C_V + 0.25} \tag{7}$$

The length of bed travel with constant velocity should be greater than the length of printing area on the cylinder. It means

$$\frac{\text{Circumference of the printing cylinder} = \pi D}{\text{Length of printing area}} > (1.5 + \sqrt{C_V + 0.25}). \tag{7_1}$$

So, the diameter D of the printing cylinder can be selected. The selection is assisted by a practical consideration: the experience indicates that D should be at least 8 inches.

Now, the equation for other elements of the design can be expressed in the term of $\frac{\pi D}{f}$.

$$\alpha = 360 \frac{\left(\frac{\pi D}{f} - 1\right)^2}{\frac{\pi D}{f}\left[2\frac{\pi D}{f} + (C_V - 2)\right]} \tag{5_1}$$

$$\beta = 360 \frac{\frac{\pi D}{f} - 1}{\frac{\pi D}{f}} \cdot \frac{\frac{\pi D}{f} + (C_V - 1)}{2\frac{\pi D}{f} + (C_V - 2)} \tag{6_1}$$

$$\gamma = 360 \frac{1}{\frac{\pi D}{f}} \tag{2_1}$$

$$S = \frac{\pi D}{C_V} \cdot \frac{\alpha}{360} \tag{1}$$

$$F = S + f$$

Due to the importance of the project, five different acceleration diagrams were checked; rather the Modified Trapezoid diagram with different values of the factor "b" were checked.

	Modified Trapezoid Acceleration – Type II				
	$b = 0; C_V = 2.0$ $C_A = 4.0$ Uniform Acceleration	$b = 0.15; C_V = 1.899$ $C_A = 4.0171$	$b = 0.25; C_V = 1.840$ $C_A = 4.0479$	$b = 0.5; C_V = 1.718$ $C_A = 4.1988$ Sine Curve Acceleration II	$b = 1; C_V = 1.571$ $C_A = 4.9348$ Cosine Curve
$\frac{\pi D}{f}$	3.0	2.96597	2.9457	2.90285	2.84937
$\alpha°$	80°	80.4536°	80.723°	81.296°	82.0046°
$\beta°$	160°	158.1697°	157.065°	154.6903°	151.6517°
$\gamma°$	120°	121.3766°	122.212°	124.0137°	126.3437°
S	$\frac{\pi D}{9} = \frac{f}{3}$	$0.1177\,\pi D =$ $= 0.349\,f$	$0.12186\,\pi D =$ $= 0.35896\,f$	$0.13144\,\pi D =$ $= 0.3815\,f$	$0.145\,\pi D =$ $= 0.41316\,f$
$F = S + f$	$\frac{4}{9}\pi D = \frac{4}{3}f$	$0.455\,\pi D =$ $= 1.349\,f$	$0.4613\,\pi D =$ $= 1.359\,f$	$0.4759\,\pi D =$ $= 1.3815\,f$	$0.4961\,\pi D =$ $= 1.41316\,f$

If it is not practical to use precise calculated angles and strokes, then it is necessary to calculate maximum accelerations.

$$A_f = C_A \,.\, S \,.\, \left(\frac{360}{\alpha}\right)^2 \quad \text{and} \quad A_r = C_A \,.\, F\left(\frac{360}{\gamma}\right)^2.$$

At speeds below 100 RPM, 3% difference may be acceptable, but at speeds above 150 RPM, it would be safer to have the difference in acceleration $\leqslant 1\%$.

EXAMPLE X

General Analysis of a Stroke with Uniform Velocity Period

Examples VI-IX show the use of uniform velocity periods under different conditions—to meet specific requirements. Sometimes, the designer has to decrease the size of a cam, due to the insufficient space for it in the machine, but the cam has to produce a certain stroke.

At the moment, it is sufficient to mention that the introduction of a uniform velocity period decreases the maximum velocity during the stroke and, as a result, the slope of the cam surface decreases; that fact permits to design a smaller cam. But, at the same time, the presence of a uniform velocity period increases the maximum value of the acceleration, thus increasing the dynamic forces required to produce the desired stroke.

A general study is offered of the influence of different periods of uniform velocity. These periods are presented as fractions of the working period and designated by the letter x. It was established that every shape of the acceleration diagram has definite values of coefficients C_V, C_A, and C_D, when equations for the velocity and for the acceleration are written for the stroke and for the working period during the action of the analyzed diagram. In the analyzed cases, a period of uniform velocity is introduced; if the equations are written for the whole stroke S and the total working period α, the values of the coefficients for the maximum velocity and for the + or — acceleration will be different and they can be compared readily with coefficients when $x = 0$.

The analysis is made for the movement of Type I with the factor $P = 1$, when the times for + and — accelerations are equal. The following nomenclature is used:

$\alpha°$ — total working period

$x\alpha°$ — uniform velocity period

$(1 - x)\alpha°$ — period of the varying velocity

T — time of the stroke S

S — total stroke

s_x — displacement with uniform velocity

$S - s_x$ — displacement with varying velocity

N	— speed of the cam in RPS	t_x	— time of uniform velocity
		$T - t_x$	— time of varying velocity

When x = 0		When x ≠ 0	
v	— max. velocity	v_x	— max. velocity = uniform velocity
A	— max. acceleration	A_x	— max. acceleration
C_V	— max. velocity coefficient	C_{VX}	— max. velocity coefficient
C_A	— max. acceleration coefficient	C_{AX}	— max. acceleration coefficient

Equations for x = 0:

$$v = C_V \cdot \frac{S}{T} = C_V \cdot S \cdot \frac{360}{\alpha} \cdot N; \qquad A = C_A \cdot \frac{S}{T^2} = C_A \cdot S \cdot \left(\frac{360}{\alpha} N\right)^2;$$

$$T = \frac{\alpha^\circ}{360N}.$$

When x ≠ 0.

The uniform velocity is equal to the maximum velocity during the ± acceleration periods. During these periods, the stroke is $(S - s_x)$ and the working period is $(1 - x) \cdot \alpha^\circ$. The coefficient C_V is the one which is characteristic for the selected acceleration diagram.

$$v_x = C_V \cdot (S - s_x) \cdot \frac{360}{(1 - x)\alpha} \cdot N \tag{a}$$

Due to a uniform rotation of the cam

$$t_x = xT = \frac{x\alpha}{360N}$$

and $$T - t_x = \frac{(1 - x)\alpha}{360N}$$

$$s_x = t_x \cdot v_x = \frac{x\alpha}{360N} \cdot C_V(S - s_x)\frac{360}{(1 - x)\alpha}N;$$

$$s_x = \frac{x}{1 - x} C_V S - \frac{x}{1 - x} C_V \cdot s_x;$$

$$s_x\left(1 + \frac{x}{1 - x} C_V\right) = \frac{x}{1 - x} C_V S;$$

$$s_X = S\frac{xC_V}{1 + x(C_V - 1)};$$

$$S - s_X = S\left(1 - \frac{xC_V}{1 + x(C_V - 1)}\right) = S\frac{1 - x}{1 + x(C_V - 1)} \tag{b}$$

Introduce $(S - s_X)$ into the equation (a).

$$v_X = C_V \cdot \frac{1}{1 + x(C_V - 1)} \cdot S\frac{360}{\alpha}N \tag{c}$$

In this expression S and $\alpha°$ represent the total stroke and the total working period, so it is possible to write

$$C_{VX} = \frac{C_V}{1 + x(C_V - 1)} \quad \text{or} \quad \frac{C_{VX}}{C_V} = \frac{1}{1 + x(C_V - 1)} \tag{1}$$

Write the equation for the maximum acceleration and introduce in it the equation (b).

$$A_X = C_A(S - s_X)\left(\frac{360}{(1 - x)\alpha}N\right)^2 = C_A \cdot \frac{1}{1 + x(C_V - 1)} \cdot \frac{1}{(1 - x)} \cdot$$

$$\cdot S\left(\frac{360}{\alpha}N\right)^2 \tag{d}$$

The equation (d) will have the familiar form if C_{AX} is introduced

$$C_{AX} = C_A\frac{1}{[1 + x(C_V - 1)] \cdot (1 - x)} \quad \text{or} \quad \frac{C_{AX}}{C_A} = \frac{1}{[1 + x(C_V - 1)](1 - x)} \tag{2}$$

High values of $x = 0.5$ and above—should be used only at low speed, due to fast growth of dynamic forces, especially in the case of Modified Sine or Sine diagrams. (See Table on page 183)

The analysis of cases when the factor $P \neq 1$ is not offered because they are very unusual and complicated for the presentation with the table—separate tables should be made for each value of P. If there is a necessity to make such an analysis, the following should be remembered.

(1) values of C_A and C_D are different;

(2) values of working periods for the acceleration and for the deceleration are different, therefore, the values of x are also different, and the analysis should be made separately for + and — acceleration.

	x =	0	0.1	0.2	0.3	0.4	0.5	0.6	0.8
$\frac{C_{VX}}{C_V}$	Sine and Mod. Trap.	1.0	0.909	0.833	0.769	0.714	0.667	0.625	0.556
	Mod. Sine	1.0	0.929	0.868	0.814	0.767	0.724	0.687	0.622
$\frac{C_{AX}}{C_A}$	Sine and Mod. Trap.	1.0	1.011	1.042	1.099	1.190	1.333	1.562	2.778
	Mod. Sine	1.0	1.033	1.085	1.163	1.278	1.478	1.717	3.109
C_{VX}	Sine and Mod. Trap.	2.0	1.818	1.667	1.538	1.429	1.333	1.250	1.111
	Mod. Sine	1.76	1.635	1.527	1.433	1.349	1.274	1.209	1.094
C_{AX}	Mod. Trapezoid	4.888	4.942	5.093	5.372	5.817	6.517	7.635	13.573
	Mod. Sine	5.528	5.710	5.998	6.429	7.065	8.013	9.492	17.189
	Sine	6.283	6.352	6.545	6.905	7.477	8.378	9.814	17.454

In the presented analysis, the working period with the varying velocity was represented by $(1 - x)\alpha$ or by $(1 - x)\frac{\alpha}{2} \cdot 2$.

In the case of $P \neq 1$, the angle $\frac{\alpha}{2}$ should be replaced in one part of the analysis by α_A and in the other by α_D; the value of x would be different because it is equal

$$\text{either to } \frac{\text{Period of varying velocity} = 2\alpha_A}{\text{Total working period} = \alpha} = x_A$$

$$\text{or to } \frac{\text{Period of varying velocity} = 2\alpha_D}{\text{Total working period} = \alpha} = x_D.$$

EXAMPLE XI

Use of Clutch-Brake Units

During recent years and, especially, after 1960 the application of clutch-brake systems has been increasing and they create additional problems for machine designers. There is a great variety in applications and, roughly, they can be divided into three major groups.

(1) Clutch-brake units are used in connection with precision indexing mechanisms which produce intermittent motions; the purpose of units is to extend dwell period by stopping driving cams and to retain fast indexing.

(2) They create intermittent motion in cases when a high precision in the positioning of moving parts is not required.

(3) They are used to start and to stop continuously running mechanisms, as conveyors, machine tools and punch presses; the latter may be connected with a number of different mechanisms, as dials, intermittently feeding devices and others.

There is a necessity to appraise a magnitude of inertial forces and torques, created by actions of clutch-brake systems—to determine stresses in transmission parts.

Usually a clutch-brake unit is installed on a shaft with a higher speed of rotation than any other shaft in the mechanism, though there are important exceptions.

Groups I and II

These groups present simpler problems; usually the clutch-brake unit is installed on a shaft of an electric motor which drives a high speed shaft of a reduction unit; a low speed shaft drives an intermittent motion mechanism, which moves a final part, as a dial or chain with fixtures for transferring from one station to another materials or parts, placed in fixtures, for some processing by different stationary tools during stops. It is reasonable to assume that friction surfaces of the clutch and of the brake are of the same quality and they are subjected to the same pressure. Then it can be expected that the acceleration created by the clutch will be lower than the deceleration caused by the brake—the speed of the motor will be somewhat decreased by the applied inertial load, while the braking surface remains stationary without any change in its positions; besides, the friction in the driven mechanism acts in the same direction as the brake.

Therefore, the dynamic forces caused by the brake will be higher and the analysis should cover primarily the load created by the brake.

The brake decelerates rotating parts of the mechanism, so only the deceleration period of the usual acceleration diagram is involved.

For a complete acceleration diagram with equal periods of + and — acceleration (P = 1) the max. value of the ± acceleration is

$$\Omega = \pm C_A \frac{S}{T^2} \qquad \text{(a)}$$

S — the stroke produced during a complete ± acceleration; since P = 1, the stroke during one half on the diagram is $\frac{S}{2}$; designate the angular movement of the shaft during the action of the brake as λ radians.

$$\lambda = \frac{S}{2} \quad \text{and} \quad S = 2\lambda$$

T — time used by a complete ± diagram; during one half of the diagram the time is $\frac{T}{2}$; designate the time of the brake action as t, then $t = \frac{T}{2}$ and $T^2 = (2t)^2$

C_A— coefficient of the maximum acceleration, which is specific for each type of acceleration diagram; it is dimensionless and has the same value for either half of the diagram; it is reasonable to expect that the braking torque increases gradually (though quickly) to its maximum and then this value remains constant until the shaft is stopped, therefore, diagrams of the Type II should be considered.

Introduce values of S and T into the equation (a)

$$\Omega = C_A \frac{2\lambda}{(2t)^2} = \frac{C_A}{2} \cdot \frac{\lambda}{t^2} \qquad (1)$$

Make a list of values of $\frac{C_A}{2}$ for four diagrams of the Type II.

Uniform acceleration diagram; $C_A = 4$; $\frac{C_A}{2} = 2.0.$

Mod. Trapezoid Diagram II; $b = 0.25$; $C_A = 4.048$; $v = 2.024$.

Sine Curve Diagram II; $b = 0.5$; $C_A = 4.2$; $v = 2.1$.

Mod. Sine Diagram II; $c = 0.25$; $C_A = 4.48$; $v = 2.24$.

For safety, accept a slightly higher value of C_A, and write the equation for max. deceleration

$$\Omega = 2.3 \frac{\lambda}{t^2} \tag{1_1}$$

Express the time t in terms of λ and the speed of the shaft N_{RPS}.

The value of λ depends on the pressure on the surfaces in contact and on the friction coefficient; it can be determined only by experiments.

Before the application of the brake, the shaft rotates with the speed which should be equal to the maximum velocity of the selected deceleration diagram.

$$w_{max} = C_V \frac{\lambda}{t} = 2\pi N.$$

The average velocity during the period of deceleration is (see page 12).

$$\frac{\lambda}{t} = \frac{w_{max}}{C_V} = \frac{2\pi N}{C_V}$$

Therefore

$$t = \frac{C_V \cdot \lambda}{2\pi N} \tag{2}$$

In the equation (1_1) the coefficient $\frac{C_A}{2}$ is selected slightly higher than for the Mod. Sine Diagram II, which has $C_V = 1.63$.

In this equation t^2 is in the denominator, so for additional safety, $C_V = 1.6$ is selected.

$$t = \frac{1.6\lambda}{2\pi N} = 0.8 \frac{\lambda}{\pi N} \tag{2_1}$$

Introduce the expression for the time t into the equation (1_1)

$$\Omega = 2.3 \frac{\lambda \pi^2 N^2}{0.64\, \lambda^2} = 35.5 \frac{N^2}{\lambda \text{ radians}} \cong 2000 \frac{N^2}{\lambda^\circ} \tag{1_2}$$

The torque on the shaft with the brake is the product of the angular acceleration and of the sum of moments of inertia of rotating parts stopped by the brake action; moments of inertia of individual parts should be converted to the shaft with the brake.

The maximum Torque $= \Omega \cdot I = 35.5 \frac{N^2}{\lambda \text{ radians}} I \cong 2000 \frac{N^2}{\lambda^\circ} \cdot I$ (3)

(see equation (2) in the Chapter I)

Rules of the Conversion of Moment of Inertia

Moment of inertia of one part $i = m \cdot r_g^2$

where $m = \text{Mass} = \frac{\text{Weight W of the rotating part}}{\text{Gravity acceleration } g = 386}$

r_g — radius of gyration, which can be found in handbooks.

In this part rotates at $N_{1\,RPS}$ and the shaft, to which the moment of inertia is converted, rotates at N_{RPS}, the converted moment of inertia is

$$i_C = i\left(\frac{N_1}{N}\right)^2 \tag{4}$$

Two special cases are presented:

(1) A chain driven by a sprocket continuously at uniform speed; the moment of inertia of the sprocket $i_S = \frac{W_S}{g} \cdot r_g^2$;

the weight of the chain is W_{CH} and the mass $m = \frac{W_{CH}}{g}$

the moment of inertia of the chain $i_{CH} = \frac{W_{CH}}{g} \cdot$ (pitch radius of the sprocket)^2;

The total converted moment of inertia

$$(i_S + i_{CH}) \cdot \left(\frac{N_S}{N}\right)^2 \tag{5}$$

N_S — the speed of the sprocket in RPS.

(2) A dial or a chain moving intermittently—with a varying velocity.

First determine the maximum angular velocity in radians per second

$$w_{max} = C_V \cdot (\text{angular stroke}) \cdot \frac{360}{\alpha^\circ} \cdot N_x = C_V \cdot \frac{2\pi}{n} \cdot \frac{360}{\alpha^\circ} \cdot N_x$$

n — number of stops per revolution of the driven part;

N_x— speed of the driving shaft of the unit in RPS;

$\alpha°$— working period of rotation of the driver, when the stroke is produced; if the driver is a continuously rotating cam, $\alpha°$ may have different values and C_v depends on the acceleration diagram selected for the intermittent motion of the driven part;

if the driver is a crank, $\alpha° = 180°$ and C_v can be taken = 1.6, for a short connecting rod, while in case of a Scotch Yoke $C_v = 1.571$. (See Chart 2, Chapter II);

The maximum angular velocity in radians/sec. should be divided by 2π, to present it in revolution per second.

$$N_y = \frac{w_{max}}{2\pi} = \frac{C_v}{n} \cdot \frac{360}{\alpha°} N_x \qquad (6)$$

Then the conversion of the moment of inertia is made by the multiplication by $\left(\frac{N_y}{N}\right)^2$; the result is entered into the sum of converted moments of inertia for the inclusion into the equation (2) for the torque T (Chapter I).

Group III

The problems may be more complicated due to a possible presence of more than one mechanism with an intermittent motion in addition to other types of movement, described in the study of Groups I and II.

In such case, it is advisable to calculate N_y for max. velocity in RPS of intermittently moving part which has the highest moment of inertia and the velocity N_x and a low n and $\alpha°$.

For other intermittently moving parts which have a lower moment of inertia and a lower velocity than in equation (6); the latter can be appraised for the moment when the major part has the maximum velocity.

A knowledge of timing of the machine and the existence of velocity tables will be helpful.

Additional calculations may be necessary for designing some transmission elements (shafts, gears, etc.) which operate only a few rotating parts.

The equation (3) for the torque should have the speed N of the transmission element under the consideration, the angle λ may be different than on the brake shaft and moments of inertia of the parts, driven by this element, should be converted (equation 4) to the speed of the transmission element, which is maybe different than the speed of the brake shaft. Very high inertia torques may be created by the action of the brake when parts with high moments of inertia are driven by a slowly rotating brake shaft; such conditions are present in punch presses when a clutch-brake is installed on the main crank shaft of the press, which has a smaller speed than the speed of the motor driving it.

When it is not reasonable to design transmission part for loads created by emergency stops and it is hopeless to expect that the angle λ of a sufficient size can be accepted by users of the equipment—then the only solution is in the introduction of a "shock absorbing" device next to the shaft with the brake.

Such a device should have a capacity to absorb a greater part of the energy stored in rotating parts and to be able to restore the proper timing between the brake shaft and the rest of the mechanism.

The energy stored in the system is

$$E = 2\pi^2 N^2 I = \frac{w^2}{2} \cdot I \qquad (7)$$

w — angular velocity of the driving shaft with the clutch-brake;
I — the sum of all moments of inertia converted to the speed of this shaft, as in equation (3), (4), (5) and (6).

The suggested analysis of the use of clutch-brake units confirms the usefulness of the coefficients C_V and C_A for a quick appraisal of acting inertial forces.

Chapter V

DESIGN DETAILS

Previous chapters covered mostly kinematics of cams and of some mechanisms, usable in automatic machines.

The offered material permits to calculate dynamic forces, necessary for different movements in machinery and to prepare data for manufacturing cams. The knowledge of these forces allows to balance them, so two or more movements employed in the machine, will be factors in limiting the speed and, therefore, the output of the machine. As a result, a proper distribution of time or of working periods of different movements is possible. This work is a part of the creation of desirable lay-out of the machine.

This chapter is devoted to different items which affect dimensions and strengths of machine details, connected with cam operated movements. In many cases, experienced designers can select all details, depending on an estimation of expected loads; and overestimation is quite usual, but it is not objectionable, as it gives a greater safety factor—providing that larger dimensions do not crowd other parts.

Generally, it is advisable to calculate dynamic forces—to be sure that they are not greater than all other forces combined; these calculations are necessary when a high speed and/or large masses are involved in the movement. In critical cases, a detailed analysis of forces, as described in the text, is justifiable.

Useful Load on Cam

It may be a combination of four or more factors.

(a) frictional resistance to the movement in the cam follower system;

(b) resistance encountered by the follower system in performing work;

(c) dynamic forces, created by the necessary acceleration of the cam follower system and, possibly, by increased vibrations;

(d) spring tension, if a spring or a similar arrangement is required to keep the cam follower in contact with the cam surface.

Items (a) and (b) are estimated by the designer. Item (c) is calculated, if dynamic properties of the cam and the mass or the moment of inertia of the cam follower system are known; item (d) can be determined only after the magnitude and the distribution of other loads is established.

Springs are not necessary, if the cam follower is in a groove or conjugate cam followers are used. If conjugate cam followers are always in contact with the cam surface, the item (d) can be eliminated. If the cam follower is in a groove and there is a chance that combined forces change the direction of their action (as during the deceleration), it is advisable to select the strength of the cam surface with an additional safety factor $1.25 \sim 1.5$—to stand the impact when the cam follower moves from one side of the groove to the other side.

When a diagram of forces (a), (b), and (c) is made, the forces, pressing the cam follower to the cam surface, are shown above a neutral line; the forces separating the follower and the cam surface are shown below the neutral line.

It is not necessary to draw a diagram precisely at every point of one cycle; maximum magnitude of forces is important and the magnitude at critical points.

The diagram on Figure V—1 is prepared for a flat cam. To decrease the spring tension, the factor P is selected $= 1.5$ for the rise of the cam and $P = 2/3$ for the drop; then the dynamic forces shown below the neutral line are smaller.

There are two important points on the diagram:

The point F, where the sum of all forces is the highest; it determines the highest pressure between the cam follower and the cam surface, which should provide the force F to be applied to the cam follower system; the actual pressure on the cam follower is greater than the force F, due to the slope of the cam surface at that point.

The point G, where the spring pressure should overcome the highest force W which tends to separate the cam follower from the cam surface. The spring tension is directly proportional to its deflection; it is possible to use different values of spring tension—q and Q—but the line connecting them should pass through the point G. It is evident that a lower force q, which represents the spring tension on a smallest cam radius, requires a higher spring tension Q on the largest cam radius. Cams with long strokes require a careful selection of springs to prevent unnecessary stresses on the maximum radius of the cam, where the spring pressure Q is applied.

It should be noted that theoretically the line presenting spring pressure is a curve because the spring tension is proportional to non-uniform movement of the cam follower system. But for the practical purpose, a straight line is acceptable.

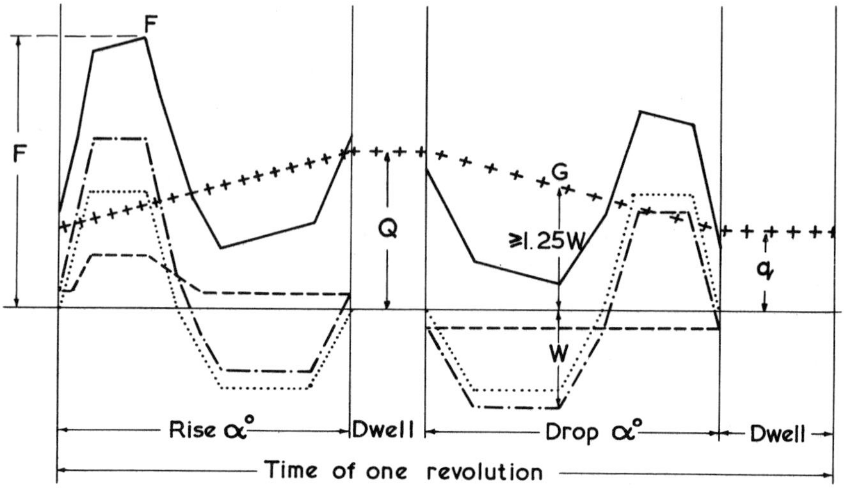

Fig. V - 1

-------- Frictional and work forces (a + b);
.... Dynamic force (c);
-·-·-·-·- Algebraic sum of forces (a + b + c)
+++++ Spring force (d);
——— Total of all forces (a + b + c + d)

The variation of the spring pressure during the stroke can be presented graphically, because a force is proportional to the deflection of the spring. Coefficient of the proportionality is called "spring factor" and is usually expressed as

K lb. per inch of deflection.

Any basic type of spring can be used—compression, extension or tortional. Tortional springs are practical mostly for smaller forces. Extension springs usually have a weak point where the spring is attached to the machine parts at both ends. Therefore, a compression spring is demonstrated, as the most practical.

Free length of the spring = l; (fig. V—2)

The length of the spring when the cam follower is on the smallest radius of the cam = l — a; "a" represents the precompression of the spring. The length of the spring when the cam follower is on the largest radius of the cam = l — (a + b). "b" may be equal to the cam stroke S or not, depending on the design.

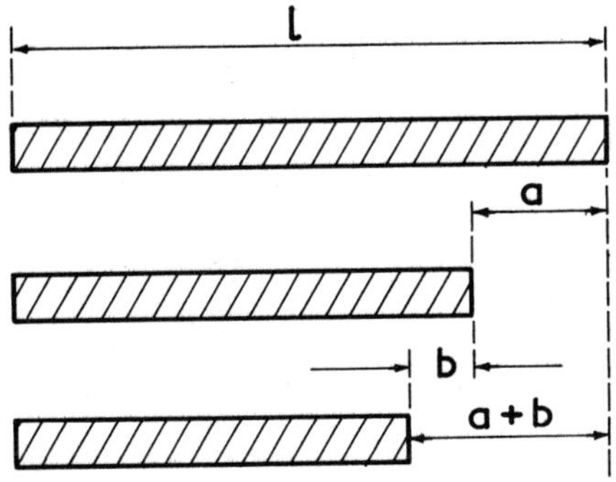

Fig. V - 2

Then the force $q = ka$;

Then the force $Q = k(a + b)$; where k is a spring factor in $\frac{lb}{inch}$;

the ratio $\frac{Q}{q} = 1 + \frac{b}{a}$ (a)

The difference between Q and q decreases when $\frac{b}{a}$ decreases; since b depends on the stroke S, the designer can only increase a, or, in other words, precompress the spring on the greater distance than the additional deflection during the stroke.

There is another way to present the influence of the ratio between a and b. Assume—the deflection b is equal the length of the stroke and the critical point G is in the middle of the stroke (actually it is very close to that). Then the required force

$$1.25\ W = q + \frac{Q - q}{2} = k\left(a + \frac{b}{2}\right) = k\,\frac{2a + b}{2}$$

$$k = 1.25\ W \cdot \frac{2}{2a + b};$$

introduce the stroke by writing

$$k = \frac{1.25\ W}{b} \cdot \frac{2}{\frac{2a}{b} + 1} = \frac{1.25\ W}{S} \cdot \frac{2}{2\frac{a}{b} + 1} \qquad \text{(b)}$$

then a table can be prepared

$\frac{a}{b}$	0.5	1.0	1.5	2	2.5	3.0
$\frac{b}{a}$	2.0	1.0	$\frac{2}{3}$	0.5	0.4	$\frac{1}{3}$
$\frac{Q}{q}$	3.0	2.0	$1\frac{2}{3}$	1.5	1.4	1.3
$k \div \frac{1.25\,W}{S} = \frac{2}{2\frac{a}{b} + 1}$	1.0	$\frac{2}{3}$	$\frac{1}{2}$	0.4	$\frac{1}{3}$	0.286

An increase of $\frac{a}{b}$ permits to use a softer spring with considerably smaller spring factor k.

The equation (b) can make a wrong impression that k is inversely proportional to the stroke S; the force W is greatly affected by the size of the stroke; therefore S appears also in the numerator; besides in the equation, according to the assumption, S introduced instead of b; in many cases $b < S$.

The selection of a proper spring may decrease the maintenance.

Actual Pressure on the Cam Follower

The diagram of forces shows the maximum useful load on the cam follower is in the direction of the movement of the follower; the maximum actual pressure on the follower is higher than the useful pressure, due to slope of the cam surface. Quite often the maximum of the useful force does not coincide with the position of the maximum slope of the cam, but it is a good practice to consider that both maxima coincide. Some forces, acting on the follower, are estimated, so a slight additional safety factor is desirable.

Cam followers may be installed with many possible directions of movement, as in the cases when the cam follower is supported by a swinging arm. The simplest case for the analysis is when the follower moves radially from the center of the cam; then the difference between the "useful" and the "actual" force depends exclusively on the slope of the cam surface. Movements in other directions, than a radial one, may increase or decrease this difference.

For a radial movement of the cam follower

$$\text{Actual pressure on the cam follower} = p = \frac{\text{Useful Pressure}}{\text{Cos } \varphi} \qquad (c)$$

where φ is the angle of the slope of the trajectory of the cam follower center.

The selection of cam dimensions is based substantially on a limit of the angle φ. Besides the value of the actual pressure on the cam follower, described by the equation (c), the appraisal of this limit depends also on the size of an allowable side pressure on the cam follower which is in the direction perpendicular to the movement of the follower.

Side pressure on the cam follower

$$p_s = U \cdot tg\varphi \qquad (d)$$

If the cam follower system slides in guides, it is proper to select $\varphi \leqslant 27°$, when $tg\varphi = 0.5$ — to decrease a possibility of binding in the guides, which may cause a crushing of the cam follower. When the possibility of binding is eliminated, it is possible to raise φ to 33° when $1/\text{Cos } \varphi = 1.2$ and $tg\varphi = 0.65$.

Higher values of the angle φ can be used for cam followers on swinging arms. During the rise on the cam, it is desirable to have the supporting arm in tension, particularly, if the follower may develop a noticeable resistance to rotation. That could happen with solid followers rotating on journal bearings or with needle bearings without cages, so the needles can skew and the bearing stops rotating. The additional friction between the follower and the cam surface may increase the pressure on the cam follower, when the resultant force—"useful" and frictional—will pass between the swivel point of the arm and the center of the cam. That does not happen when the swinging arm is in tension.

Usually, the maximum pressure U does not coincide with φ max., so for preliminary considerations, it is safe to accept

$$p = 1.2\ U_{max} \qquad (e)$$

The table shows the influence of the angle φ on pressure between the follower and the cam surface.

φ	21°	24°	27°	30°	33°
Total Pressure $\frac{p}{U}$	1.07	1.095	1.112	1.155	1.195
Side Pressure $\frac{p_s}{U}$	0.38	0.45	0.51	0.58	0.65

φ	36°	39°	42°	45°	48°
Total Pressure $\frac{p}{U}$	1.236	1.287	1.347	1.414	1.495
Side Pressure $\frac{p_s}{U}$	0.73	0.81	0.9	1.0	1.11

Selection of Cam Followers

The size of the follower is affected by the maximum actual pressure p; by the speed of rotation and, in the case of antifriction bearings, by the expected life.

Antifriction bearings, used as cam followers, have advantages over solid rollers rotating on journal bearings:

(a) lower friction coefficient with less chance of sliding;

(b) more dependable lubrication, which reduces maintenance;

(c) high surface hardness, so the designer has to provide only a sufficient hardness of the cam surface;

(d) availability of standard sizes and, therefore, easy replacements.

On modern cams with controlled accelerations, ball bearings can be used successfully at high speed, unless there is a shock load from the work performed by the cam follower systems. For higher loads and in the case of shock load, needle and roller bearings should be considered.

The Anti-Friction Bearing Manufacturers' Association established rules for the calculation for each bearing the limiting capacities:

Basic Dynamic Capacity — C and
Static Capacity — C_0.

Eventually, all manufacturers of standard antifriction bearings will use this method, so the designer can depend on the published dynamic capacity of each bearing.

The influence of the speed of a bearing and of the expected life on the allowable load was established by numerous tests. Therefore, it is possible to prepare tables of correction factors for the capacity C.

The basic equation used for making the table is

$$\text{Allowable Load } p_c = \frac{\text{Basic Dynamic Capacity C}}{\text{Total Factor F}} \cdot f_h \qquad (1)$$

The allowable load p_c determined from this equation should not be greater than

$$0.4 \, . \text{ Static Capacity } C_0 \qquad (2)$$

The tables are prepared for 8000 hrs. life of 90% of a group of antifriction bearings used as cam followers in production machinery. In special cases when the bearing is designed for a life of "H" hours instead of 8000 hrs., the data from the tables should be multiplied by the factor.

$$f_h = \sqrt[3]{\frac{8000}{H}} \qquad (3)$$

The table shows the factor f_h as a function of expected life H hours.

H hrs Life	10	50	100	200	500	1000
Factor f_h	28.3	5.43	4.31	3.42	2.52	2.0

H hrs Life	2000	4000	6000	8000	12,000	16,000
Factor f_h	1.59	1.26	1.1	1.0	0.874	0.794

When this factor is used, still the rule of the expression (2) should be observed.

The total factor F for 8000 hrs. life is affected by the speed of the bearings and, in the case of cam followers, by the rotation of their outer races.

The factor F is higher for cam followers, its value is obtained by multiplying the factor F for bearings with the stationary outer races by one of the following factors.

Needle bearings	1.2
Extra light ball bearings	1.3
Light ball bearings	1.5
Medium ball bearings	1.7
Heavy ball bearings	1.9

For roller bearings, the correction can be selected from this list for the same ratio between the diameter of rollers and the diameter of their pitch line; approximately, the correction factor can be selected between 1.4 and 1.6; or equal to light ball bearings.

	Bearing Speed RPM	100	200	400	600	750	1000	1500	2000	3000	4000	6000	8000
	Stationary Outer Race	3.7	4.57	5.75	6.6	7.11	7.85	8.96	9.81	11.28	12.4	14.22	15.63
Factor F	Cam Followers: Needle bearings	4.44	5.48	6.90	7.92	8.53	9.42	10.75	11.77	13.54	14.88	17.06	18.76
	Extra light ball bearings	4.81	5.94	7.48	8.58	9.24	10.21	11.65	12.75	14.66	16.12	18.49	20.32
	Light ball bearings	5.55	6.86	8.63	9.90	10.67	11.78	13.44	14.72	16.92	18.60	21.33	23.45
	Med. ball bearings	6.29	7.77	9.78	11.22	12.09	13.35	15.23	16.68	19.18	21.08	24.17	26.57
	Heavy ball bearings	7.03	8.68	10.93	12.54	13.51	14.92	17.02	18.64	21.43	23.56	27.02	29.7

	Bearing Speed RPM	10,000	15,000	20,000	25,000	30,000	40,000
	Stationary Outer Race	16.88	19.3	21.3	22.95	24.25	26.75
Factor F	Cam Followers: Needle bearings	20.26	23.16	25.56	27.54	29.10	32.1
	Extra light ball bearings	21.94	25.09	27.69	29.84	31.53	34.78
	Light ball bearings	25.32	28.95	31.95	34.43	36.38	40.13
	Medium ball bearings	28.7	32.81	36.21	39.02	41.23	45.48
	Heavy ball bearings	32.07	36.67	40.47	43.61	46.08	50.83

The cam should have a slightly wider face than the width of the selected standard cam follower ($^1\!/_{16}$ inch +). The hole in the bearing determines the diameter of the stud and its strength should be checked, especially in case of cantilever design for cam followers operating in grooves. If the load requires undesirable high hardness of the cam surface, antifriction bearings can be installed with rings around them, which are wider than the face of the bearing; the hardness of the ring surface should be the same or higher than the hardness of the cam surface. An increase of the diameter of the cam follower decreases stresses in the area of contact between the follower and the cam.

Strength of the Cam Surface

It is influenced by three basic factors:

(1) properties of the cam material;

(2) dimensions of the cam follower — the radius and the width;

(3) radius of curvature of the cam surface—concave or convex.

Plate cams have a concave surface near the smallest radius of the cam and a convex surface near its maximum radius. Face cams and any cams with grooves have at each point—one side is concave and the other is convex; it is important along which side the cam follower has its contact with the side of the groove.

The allowable load on the cam surface is based on allowable compressive stresses in the area of contact. It can be calculated from the equations developed by H. Hertz and applied to the case of two cylindrical surfaces in contact under load; the allowable load is calculated in lb. per inch of width of the contact area.

$$P_1 = \left(\frac{\delta}{F}\right)^2 \cdot \frac{r}{1 \pm \dfrac{r}{\rho_c}} \qquad (1)$$

or $$\delta = F\sqrt{P_1 \cdot \frac{1 \pm \dfrac{r}{\rho_c}}{r}} \qquad (1_1)$$

P_1 — allowable load — lb. per inch of width;

δ — allowable compressive stress — psi;

r — radius of the cam follower — inch;

ρ_c — radius of curvature of the cam surface — inch;

F — factor affected by the properties of materials in contact; it is presented by the equation (2).

+ sign is used for convex cam surface and — sign for concave surface.

The allowable load δ is greater for a concave surface and it is smaller for the convex surface, so the latter is more important.

$$F = \sqrt{\frac{E_1 E_2}{\pi[(1-\mu_2^2)\,E_1 + (1-\mu_1)E_2]}} \qquad (2)$$

E_1 — modulus of elasticity of cam material — psi;

E_2 — modulus of elasticity of roller material — psi;

μ_1 — Poisson's ratio for cam material;

μ_2 — Poisson's ratio for roller material.

The values of F were calculated for antifriction bearings with heat-treated steel surface and for different cam materials; or for $E_2 =$ 30, 000, 000 psi and $\mu = 0.3$.

Cam material:

Steel	$E_1 = 30 \times 10^6$ psi;	$\mu_1 = 0.3$;	$F = 2300$;
Cast Steel	$E_1 = 28 \times 10^6$ psi;	$\mu_1 = 0.265$;	$F = 2250$;
Stainless Steel	$E_1 = 27 \times 10^6$ psi;	$\mu_1 = 0.305$;	$F = 2250$;
Cast Iron	$E_1 = 17 \times 10^6$ psi;	$\mu_1 = 0.25$;	$F = 1950$;
Bronze	$E_1 = 14.5 \times 10^6$ psi;	$\mu_1 = 0.355$;	$F = 1900$;
Aluminum Alloys	$E_1 = 10 \times 10^6$ psi;	$\mu_1 = 0.33$;	$F = 1650$.

This list is made for average values of E_1, and μ_1 and in specific cases, the value of F should be recalculated.

If plastic cam followers are used, the values of E_2 and μ_2 are changed; then F will be considerably lower; rollers made of Delrin and installed on needle bearings may compete with light ball bearings; but they can be used only without springs—as conjugate followers or in a groove. The reason for that is in the formation of flats even under a light pressure from springs when the mechanism is not in operation.

The pressure of a roller on a curved surface creates also shearing stresses. They approximately $= 0.3\delta_{max}$, so they are not as important as the compressive stresses, except near the edges of the cam; that is the reason for having the face of the cam wider than the face of the cam follower.

Since the shearing stresses are mentioned, it is proper to give the equation representing them

$$\tau = 0.175\sqrt{\frac{E_1 E_2}{E_1 + E_2}} \cdot \sqrt{P_1 \cdot \frac{1 + \frac{r}{\rho_c}}{r}} = G \cdot \sqrt{P_1 \frac{1 + \frac{r}{\rho_c}}{r}} \qquad (3)$$

The maximum τ_{max} are located on the depth $\cong 0.4b$, where b is the width of the area of the contact (the length of this area is 1 inch).

$$b = 2.15\sqrt{\frac{E_1 + E_2}{E_1 E_2}}, \sqrt{P_1 \frac{r}{1 + \frac{r}{\rho_c}}} = H\sqrt{P_1 \cdot \frac{r}{1 + \frac{r}{\rho_c}}} \qquad (4)$$

$$G = 0.175\sqrt{\frac{E_1 E_2}{E_1 + E_2}} \qquad (3_1)$$

$$H = 2.15\sqrt{\frac{E_1 + E_2}{E_1 E_2}} \qquad (4_1)$$

When both contacting surfaces are convex, the maximum compressive stress described by the equation (1_1), and the maximum shearing stress τ_{max} from the equation (3) are affected in a similar way by the factor

$$\sqrt{\frac{1 + \frac{r}{\rho_c}}{r}}$$

It is highly desirable for a designer to appraise the necessary strength of the cam surface without lengthy calculation, when the force P_1 is determined after the size of the cam follower is selected. For that purpose, a chart 24 is presented giving the allowable force P_1 for different hardness of a flat surface which has the radius of curvature $\rho_c = \sim$ and the ratio $\frac{r}{\rho_c} = 0$. Then from the equation (1)

$$P_1 = \left(\frac{\delta}{F}\right)^2 \cdot r \qquad (5)$$

The chart shows the allowable load P_1 for different radii r and separate lines are drawn for different hardness on Rockwell scale "C". When the hardness is RC-60, it is possible to select the allowable stress $\sigma =$ 0.5 × (ultimate tensile strength) or with the safety factor 2, because the yield strength is close to the ultimate.

For lower values of the hardness on the Rockwell scale C (below RC-45), arbitrary increased safety factors are used:

RC-40 — σ = 86, 500 psi;
RC-30 — σ = 59, 000 psi;
RC-20 — σ = 34, 500 psi;

When the maximum load on the cam occurs where the cam surface is concave, this chart gives conservatively estimated stresses. For the convex part of the cam surface, the chart data should be corrected—the allowable load P_1 should be lowered, or the actual calculated load per inch of cam follower width should be increased, as shown in the equation (6).

$$\frac{\text{Expected load on the roller}}{\text{Width of the cam follower}} \cdot \left(1 + \frac{r}{\rho_c}\right) = \left(\frac{\sigma}{F}\right)^2 r \qquad (6)$$

—(from the Chart 24)

While the allowed load should be increased by multiplying it by $1 + \frac{r}{\rho_c}$, the compressive stresses will be increased by $\sqrt{1 + \frac{r}{\rho_c}}$ times.

The Table gives an idea of the influence of the radius of curvature of the cam surface on the allowable load on a convex surface in comparison to the load with the allowable load on a flat surface.

It is presented as a ratio

$$\frac{\text{Allowable load on convex surface}}{\text{Allowable load on flat surface}} = \frac{1}{1 + \frac{r}{\rho_c}}$$

The stresses are presented by a similar ratio for the same load

$$\frac{\text{Stresses } \sigma \text{ on convex surface}}{\text{Stresses } \sigma \text{ on flat surface}} = \sqrt{1 + \frac{r}{\rho_c}}$$

Radius of convex curvature in terms of the cam follower radius ρ_c/r	Flat ~	9	4	3	2	1.5	1.0	0.75	0.5	0.25
Increase of the actual force for use with the chart to select the hardness of the cam $1 + \frac{r}{\rho_c}$	1.0	1.111	1.25	1.333	1.5	1.667	2.0	2.333	3.0	5.0
$\frac{\text{Allow. load on convex surface}}{\text{Allow. load on flat surface}} = \frac{1}{1 + \frac{r}{\rho_c}}$	1.0	0.9	0.8	0.75	0.667	0.6	0.5	0.4286	0.333	0.2
$\frac{\text{Stress } \sigma \text{ on convex surface}}{\text{Stress } \sigma \text{ on flat surface}} = \sqrt{1 + \frac{r}{\rho_c}}$	1.0	1.054	1.118	1.155	1.225	1.291	1.414	1.527	1.732	2.236
Increases of stresses under the same load in %	0.0	5.4%	11.8%	15.5%	22.5%	29.1%	41.4%	52.7%	73.2%	123.6%

The stresses increase fast when the convex curvature ρ_c decreases below 2r; it is desirable to keep $\rho_c > 2r$ or, at least, $\rho_c > 1.5$.

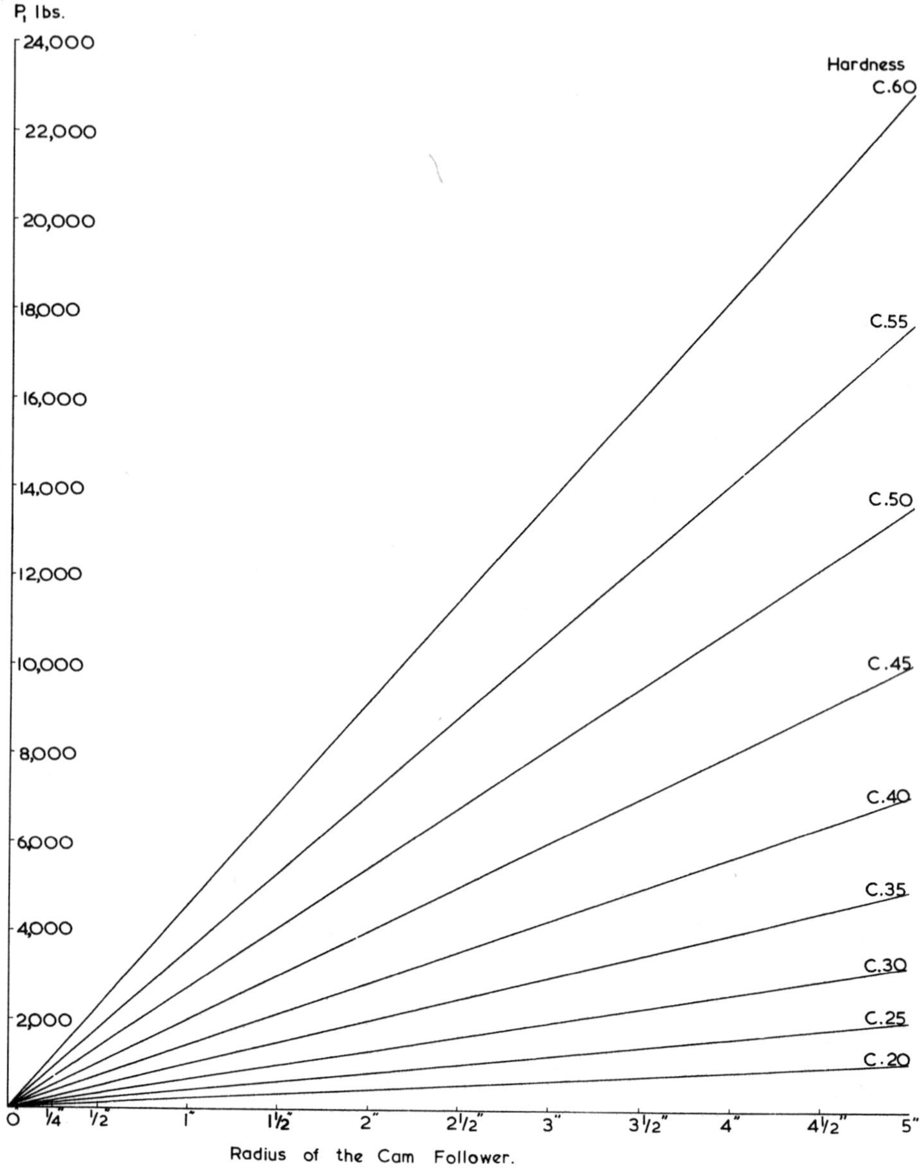

Chart 24. Allowable Pressure on Cam Follower on a Flat Surface

Calculation of the Slope of the Cam

The problem is more complicated for a flat cam, due to a continuous change of radii to the cam surface; barrel cams have a constant radius, therefore, the solution is considerably easier. To simplify the analysis, it is assumed that the cam follower moves radially on a straight line.

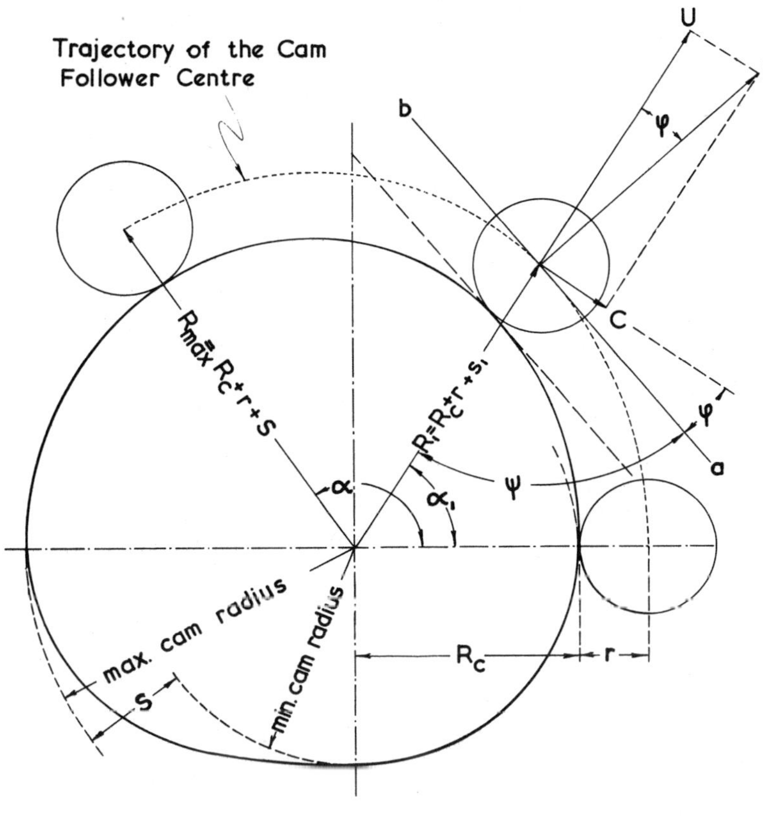

Fig. V - 3

R_c — min. radius of the cam

r — radius of the cam follower

$R_{min} = R_c + r$ — min. distance between centers of the cam and the follower

S — stroke of the cam follower

$R_{max} = R_c + r + S$ — max. distance between centers of the cam and follower

$R_1 = R_c + r + s_1$—distance between centers when the cam turns on α_1°, where

s_1 — displacement of the follower at that moment

α° — working period when the stroke S is produced

α_1° — working period when the follower moves on a distance, s_1

ab — tangent to the trajectory of the cam follower center at the angle of cam rotation α_1°

P_1 — force acting on the follower in the direction perpendicular to the tangent ab

U — useful force acting in the direction of the movement of the cam follower

C — component of the force P which determines the torque on the cam shaft

N — uniform speed of the cam in RPS

T — $\frac{\alpha^\circ}{360N}$ -time required to move the cam follower on the distance S

t — variable time, which changes from 0 to T during the full stroke.

Values of r, α°, S and U are already known.

Minimum radius of the cam, R_c, should be determined by the designer with due consideration of the maximum allowable slope of the cam surface. s_1 is the radial displacement of the cam follower when the cam rotates on the angle α_1°, from the beginning of the stroke. It is assumed that the cam rotates at a uniform speed; therefore, the angle α_1 changes from 0 to α in the same way as the time t changes from 0 to T, so the ratio $\frac{\alpha_1}{\alpha} = \frac{t}{T}$, and

$$\alpha_1 = \alpha \frac{t}{T} \qquad \text{(a)}$$

The displacement s_1 is also a function of the time t. Since the designer can select a known acceleration diagram, s_1 can be presented as

$$s_1 = S \cdot f_S\left(\frac{t}{T}\right) \qquad \text{(b)}$$

where $f_S\left(\frac{t}{T}\right)$ varies between 0 and 1.

Then the vector R_1 can be written as

$$R_1 = R_C + r + S \cdot f_S\left(\frac{t}{T}\right) \qquad \text{(c)}$$

At that moment, the tangent of the angle φ can be presented as a ratio of infinitely small movements.

$$\tan \varphi = $$

$$= \frac{\text{Change of the length of the vector } R_1 \text{ or of the displacement } s_1}{\text{Movement along the trajectory of the cam follower center}}$$

$$= \frac{d(R_1)}{R_1 \cdot d(\alpha_1)} \qquad \text{(d)}$$

From the equation (c)

$$d(R_1) = S \cdot d\left[f_S\left(\frac{t}{T}\right)\right] \qquad \text{(e)}$$

From the equation (a)

$$d(\alpha_1) = \alpha \cdot d\frac{t}{T},$$

if the angle α is in radians or

$$d(\alpha_1) = \alpha^\circ \frac{\pi}{180} \cdot d\left(\frac{t}{T}\right) \qquad \text{(f)}$$

Introduce the expressions (c), (e) and (f) into the equation (d).

$$\operatorname{Tan} \varphi = \frac{S \cdot d\left[f_S\left(\frac{t}{T}\right)\right]}{\left[R_C + r + S \cdot f_S\left(\frac{t}{T}\right)\right] \frac{\alpha^\circ \pi}{180} \cdot d\left(\frac{t}{T}\right)} = \frac{\dfrac{\alpha\left[f_S\left(\frac{t}{T}\right)\right]}{d\left(\frac{t}{T}\right)} \cdot \dfrac{180}{\alpha^\circ \pi}}{\dfrac{R_C + c}{S} + f_S\left(\frac{t}{T}\right)}$$

$$\frac{d\left[f_S\left(\frac{t}{T}\right)\right]}{d\left(\frac{t}{T}\right)} = f_S^1\left(\frac{t}{T}\right) = C_V \cdot f_V\left(\frac{t}{T}\right)$$

which is the variable dimensionless part of the velocity equation with the max. velocity coefficient

Then

$$\tan \varphi = \frac{180}{\alpha^\circ \pi} \cdot \frac{C_V \cdot f_V\left(\frac{t}{T}\right)}{\frac{R_c + r}{S} + f_S\left(\frac{t}{T}\right)} \tag{1}$$

When the acceleration diagram is known, the dimensionless equations for the velocity and of the displacement— $C_V \cdot f_V\left(\frac{t}{T}\right)$ and $f_S\left(\frac{t}{T}\right)$ —can be written into the expression for tan φ. Then it is possible to find conditions for tan φ_{max}.

By changing values of α and $\frac{R_c + r}{S}$, it is possible to find values of φ_{max} for many different conditions. The data can be presented on the chart with φ°_{max} plotted on the abscissa line and $\frac{R_c + r}{S}$ as ordinates with angles α° as parameters. This chart shows a certain working period α° connections between the maximum slope angle and the value of $\frac{R_c + r}{S}$, where only R_c—the smallest radius of the cam—is not known and can be determined.

$$R_c = S \cdot (\text{ratio} \frac{R_c + r}{S} \text{ from the chart}) - r \tag{2}$$

Such charts, 25, 26, 27 and 28, are offered for four acceleration diagrams: Modified Trapezoid Diagrams I and III, both with $P = 1.5$ (or $2/3$) as suitable for use for flat cams with springs holding the cam followers. Modified Sine Diagrams I and III, both with $P = 1$; suitable for movements of cam follower without springs holding the follower against the cam surface.

The diagrams for φ_{max} are drawn on the left sides of these charts, while on the right side of the data for the cam curvature is plotted; both sides have a common ordinate.

Slope on Barrel Cams

The analysis is simple because the radius R of the cam is constant. R represents the radius from the center of the cam to the nearest edge of the cam follower because in that region the slope is steeper than at any place along the face of the cam follower.

Fig. V—4 represents the open (unrolled) circumference of the cam and a greatly enlarged detail of one point; the curved line represents the trajectory of the center of the cam follower.

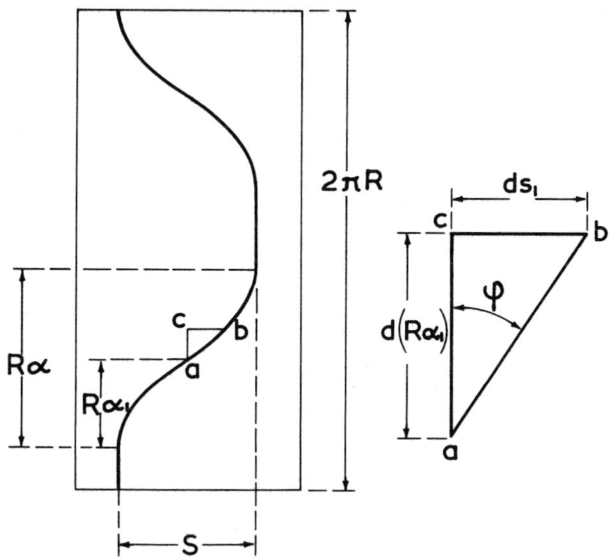

Fig. V - 4

α — working period in radians;

α° — working period in degrees;

S — total stroke;

s_1 — varying stroke $= S \cdot f_S\left(\frac{t}{T}\right)$;

α_1 and α° — variable angle of rotation;

$$\alpha_1 = \alpha\frac{t}{T};\ d(\alpha_1) = \alpha \cdot d\left(\frac{t}{T}\right) = \frac{\alpha^\circ\pi}{180} \cdot d\left(\frac{t}{T}\right);$$

$$\tan\varphi = \frac{ds_1}{d(R\alpha_1)} = \frac{S \cdot d\left[f_S\left(\frac{t}{T}\right)\right]}{R\alpha^\circ\frac{\pi}{180} \cdot d\left(\frac{t}{T}\right)} = \frac{S}{R} \cdot \frac{180}{\alpha^\circ\pi} \cdot C_V \cdot f_V\left(\frac{t}{T}\right) \qquad (3)$$

It is known that the max. of $f_V\left(\frac{t}{T}\right) = 1$, so

$$\tan \varphi_{max} = C_V \cdot \frac{S}{R} \cdot \frac{180}{\alpha^\circ \pi} \tag{4}$$

The minimum size of the cam for the same slope φ_{max} can be obtained by using an acceleration diagram with a smaller coefficient C_V, as Modified Sine diagram with $c = 0.25$. When a limited space is encountered, it is possible to decrease the coefficient C_V by the introduction of a uniform velocity period in the middle of the stroke. The example X in the Chapter IV supplies a necessary information about the influence of the uniform velocity period on the coefficient C_{VX}; also, the table shows an increase of the maximum acceleration coefficient C_{Ax}. Generally speaking, the use of uniform velocity period cannot be recommended for cams operating at a speed in hundreds RPM.

On page 203 the table shows the unfavorable conditions when stresses increase where the cam surface has a convex shape. Usually, dynamic forces are sufficiently great, so during the deceleration period, when the convex shape appears, the cam follower moves to the other side of the groove, which is concave and can stand a greater pressure. Therefore, for barrel cams with a groove as well as for face cams with the groove, the strength of the cam surface can be taken from the Chart 24 for rollers running on a flat surface. The same is true for any system of conjugate cam followers.

Calculation of the Curvature of the Cam Surface

Equations for the displacements are written for the cam follower center. The use of them in the analysis of the curvature will give instantaneous radii or the curvature of the trajectory of this center. The curvature of the cam surface can be easily derived from that. On the concave part of the cam surface, the centers of its curvature are outside the cam body (in case of flat cams) and on the convex part, these centers of curvature are inside of the cam body.

Designate radii of curvature of the trajectory of the cam follower center as ρ_t and the corresponding radii of the cam surface as ρ_c. Then on the concave part of the cam, the radii of the curvature of the cam surface will be

$$\rho_c = \rho_t + r \tag{a}$$

and on the convex part

$$\rho_c = \rho_t - r \tag{b}$$

where r is the radius of the cam follower.

Minimum allowable radii ρ_c of the cam surface curvature can be set by reasoning:
(1) on the concave part of the cam ρ_c should be greater than the radius r of the cam follower—to prevent the undercutting of the cam surface during the manufacturing process; therefore, the requirement is $\rho_c > r$ and from the equation (a)

$$\rho_t > 0 \quad \text{and} \quad \frac{\rho_c}{r} > 1 \tag{c}$$

(2) on the convex part of the cam ρ_c should not be so small that the stresses in the cam surface will be excessive; the Table of Chapter V (page 203) shows the stresses increase faster when ρ_c is smaller than 2r—when the stresses are 22.5% higher than in the case of a cam follower on a flat surface. So according to the equation (b), it is desirable to have

$$\rho_t \geqslant 3r \quad \text{or} \quad \frac{\rho_t}{r} \geqslant 3 \tag{d}$$

If the same nomenclature is used, as for the analysis of the slope of the cam surface (page 206), the basic equation of the radius of curvature of the trajectory of the cam follower center is

$$\rho_t = \frac{\left[R_1^2 + \left(\frac{dR_1}{d\alpha_1}\right)^2\right]^{3/2}}{R_1^2 + 2\left(\frac{dr_1}{d\alpha_1}\right)^2 - R_1 \cdot \frac{d^2R_1}{d\alpha_1^2}}$$

$R_1 - R_c + r + s_1$ represents a radial distance from the center of the cam to the trajectory at an angle α_1. $R_c + r$ is constant, so it disappears in the derivatives and the equation will take the form

$$\rho_t = \frac{\left[(R_c + r + s_1)^2 + \left(\frac{ds_1}{d\alpha_1}\right)^2\right]^{3/2}}{(R_c + r + s_1)^2 + 2\left(\frac{ds_1}{d\alpha_1}\right)^2 - (R_c + r + s_1) \cdot \frac{d^2s_1}{d\alpha_1^2}} \tag{1}$$

The derivatives are shown with the variable angle α_1, while the equations for designing cams are written with the variable time t or the variable ratio $\frac{t}{T}$.

Previously, it was established

$$\frac{\alpha_1}{\alpha} = \frac{t}{T} \quad \text{and} \quad t = \frac{T}{\alpha} \cdot \alpha_1$$

then

$$dt = \frac{T}{\alpha} \cdot d\alpha_1 \tag{e}$$

$$(dt)^2 = \frac{T^2}{\alpha^2} \cdot (d\alpha_1)^2 \tag{f}$$

The equations for displacement, velocity and acceleration, can be written

$$s_1 = S \cdot f_S\left(\frac{t}{T}\right) = S \cdot f_S\left(\frac{\alpha_1}{\alpha}\right);$$

$$\frac{ds_1}{dt} = \frac{S}{T} \cdot C_V \cdot f_V\left(\frac{t}{T}\right) = \frac{S}{T} \cdot C_V \cdot f_V\left(\frac{\alpha_1}{\alpha}\right)$$

$$\frac{d^2s_1}{dt^2} = \frac{S}{T^2} \cdot C_A \cdot f_A\left(\frac{t}{T}\right) = \frac{S}{T^2} \cdot C_A \cdot f_A\left(\frac{\alpha_1}{\alpha}\right)$$

Either form is acceptable.

Introduce in the left side dt and dt^2 from the equations (e) and (f); transfer $\frac{T}{\alpha}$ and $\frac{T^2}{\alpha^2}$ to the right side. Then these equations will have the form acceptable for the equation (1).

$$s_1 = S \cdot f_S\left(\frac{t}{T}\right);$$

$$\frac{ds_1}{dt} = \frac{S}{\alpha} \cdot C_V \cdot f_V\left(\frac{t}{T}\right);$$

$$\frac{d^2s_1}{dt^2} = \frac{S}{\alpha^2} \cdot C_A \cdot f_A\left(\frac{t}{T}\right).$$

Introduce them into the equation (1)

$$\rho_t = \frac{\left\{\left[R_c + r + S \cdot f_S\left(\frac{t}{T}\right)\right]^2 + \left[\frac{S}{\alpha} \cdot C_V \cdot f_V\left(\frac{t}{T}\right)\right]^2\right\}^{3/2}}{\left[R_c + r + S \cdot f_S\left(\frac{t}{T}\right)\right]^2 + 2\left[\frac{S}{\alpha} \cdot C_V \cdot f_V\left(\frac{t}{T}\right)\right]^2 - \left[R_c + r + S \cdot f_S\left(\frac{t}{T}\right)\right] \frac{S}{\alpha^2} \cdot C_A \cdot f_A\left(\frac{t}{T}\right)} \tag{2}$$

The minimum limits for ρ_t are shown in equations (c) and (d), expressed in terms of the cam follower radius r. To do that with the equation (2), the following method was used.

Introduce $R_c + r = xr$ (g)

and $S = yr$ (h)

After transformations, the equation (2) can be written in many different ways; the following form is selected—it represents the form of limits in equations (c) and (d).

$$\frac{\rho_t}{r} = \frac{\left\{\left[\alpha x + \alpha y \cdot f_s\left(\frac{t}{T}\right)\right]^2 + y^2\left[C_V \cdot f_V\left(\frac{t}{T}\right)\right]^2\right\}^{3/2}}{\alpha\left[\alpha x + \alpha y \cdot f_s\left(\frac{t}{T}\right)\right]^2 + 2\alpha y^2\left[C_V \cdot f_V\left(\frac{t}{T}\right)\right]^2 - \left[\alpha x + \alpha y \cdot f_s\left(\frac{t}{T}\right)\right] \cdot yC_A \cdot f_A\left(\frac{t}{T}\right)} \quad (3)$$

From equations (g) and (h)

$$\frac{x}{y} = \frac{R_c + r}{S} \quad (4)$$

This ratio enters also into the equation for tan φ and it is used on Charts 25, 26, 27 and 28 as ordinates, so it becomes possible to use the same ordinates for the allowable curvature, as for diagrams for φ_{max}.

In such a case, the value of

$$y = \frac{S}{r}$$

can be used on abscissa line, because both—S and r—are already selected.

As a result, the curves on the left side from the vertical scale (ordinates) permit to connect the size of the cam R_c with the maximum slope of the cam and the curves to the right from the vertical scale, drawn also with the angles α as parameters, allow to see if the radius of the curvature is equal or above the desirable minimum.

The equation (3) is complicated. It has components which describe properties of the selected acceleration diagram and the known values of α and y. So, the problem is narrowed for finding values of x which sets the minimum limit for the ratio $\frac{\rho_t}{r}$.

To find the relations between this ratio and x for numerous combinations of α and y, for the most usable acceleration diagrams and for the whole stroke would be a long and tedious work. However, it is possible to guess at what point of the stroke the minimum radius of curvature can be expected and make calculations in the region near that point. This approach may give results which are a few percentage points off the precise answers, but this fact is tolerable due to many estimations which are practically unavoidable in the designing of cams: as the selection of some loads, increase of stresses due to vibrations, graphical appraisal of maximum slope of the cam surface, and even the strength of the cam surface.

These considerations also lead to the decision to make calculations for α expressed in simple figures in radians, as 0. 5; 1. 0 and 2 and mark in degrees on the charts as 30°, 60°, and 120°, respectively.

The purpose of the charts is to give the designer a chance to check quickly the slope and the curvature for a selected minimum radius of the cam or vice versa; at the same time to provide answers with a precision sufficient for the practical use.

Curvature of the Concave Part of the Cam

To prevent the undercutting of the cam surface ρ_t should be greater than zero (equation (c)).

(1) During the dwell on the smallest radius R_c the radius of the curvature ρ_t is positive and equal $R_c + r$, which is evident. This result can be received also from the equation (2), when the displacement $f_S\left(\frac{t}{T}\right)$, the velocity $f_V\left(\frac{t}{T}\right)$ and the acceleration $f_A\left(\frac{t}{T}\right)$ of the cam follower are zero. When the cam follower enters the rise on the cam, the numerator of the equation (2) increases due to increases in displacements $f_S\left(\frac{t}{T}\right)$ and in the velocity $f_V\left(\frac{t}{T}\right)$. In the denominator, the acceleration, represented by $f_A\left(\frac{t}{T}\right)$, increases faster than two other items; since it enters with minus sign, the denominator decreases and soon becomes equal zero and ρ_t becomes infinity, which means a flat surface. Further increase of the acceleration makes the denominator negative and the surface becomes concave. But the numerator also continues to grow, so ρ_t cannot become equal zero and the undercutting cannot happen.

(2) A different condition exists when the cam produces the Type III movement and the change between the forward and return strokes (without a dwell between them) happens at the minimum radius R_c of the cam (it means the cam has a dwell on the maximum radius of the cam). The critical point is at the R_c, where the displacement $f_S\left(\frac{t}{T}\right)$ and the velocity $f_V\left(\frac{t}{T}\right)$ come to zero, but the acceleration $f_A\left(\frac{t}{T}\right)$ does not.

Then the equation (3) will have the form

$$\frac{\rho_t}{r} = \frac{(\alpha x)^3}{\alpha(\alpha x)^2 - \alpha xyC_A} = \frac{(\alpha x)^2}{\alpha^2 x - yC_A} = \frac{x^2}{x - \frac{y}{\alpha^2}C_A} \tag{5}$$

When the working period α is small and $y = \frac{S}{r}$ is large, the denominator is negative and it can be large. But the numerator $= (R_c + r)^2$ is not zero, therefore, ρ_t is always greater than 0. Therefore, the possibility of undercutting a concave part is not the factor which controls the minimum radius R_c of the cam.

Curvature of the Convex Part of the Cam

The minimum ratio $\frac{\rho_t}{r} \geqslant 3$ was offered in the equation (d). The cam surface is convex during the deceleration period of the rise on the cam follower, because $f_A\left(\frac{t}{T}\right)$ is negative and the equations (2) or (3) have only positive components in the denominator. A high magnitude of $f_A\left(\frac{t}{T}\right)$ increases the denominator. When the deceleration still has its maximum value, the velocity is substantially lower than its maximum value, so the numerator approaches some lower value. The actual minimum value of ρ_t should be somewhere close to this point.

(1) For the movements of Type I, the analyzed point was selected at $0.875\alpha°$. The values of $f_S\left(\frac{t}{T}\right)$; $C_V \cdot f_V\left(\frac{t}{T}\right)$ and $C_D \cdot f_A\left(\frac{t}{T}\right)$ were taken for this point. Then for different combinations of values of α and y the values of x were found for $\frac{\rho_t}{r} = 3$. Finally, the ratios $\frac{x}{y} = \frac{R_c + r}{S}$ were determined and the results are plotted on Charts 25 and 26 together with diagrams for maximum slopes of the cam surface.

(2) In cases with movements of the Type III, the convex surface of the cam appears near the maximum radius of the cam, when forward and return strokes meet without a dwell between them. At that point, the deceleration has its maximum value, while the displacement $f_S\left(\frac{t}{T}\right) = 1$ and the velocity of the cam follower are zero. Then the equation (3) is greatly simplified.

$$\frac{\rho_t}{r} = \frac{(\alpha x + \alpha y)^3}{\alpha(\alpha x + \alpha y)^2 + (\alpha x + \alpha y)C_D} = \frac{(x + y)^2}{(x + y) + \frac{y}{\alpha^2}C_D}$$

This equation can be solved for x.

$$x = 0.5\left[\frac{\rho_t}{r} - 2y + \sqrt{\left(\frac{\rho_t}{r}\right)^2 + 4\frac{y}{\alpha^2}C_D}\right] \tag{6}$$

+ sign should be used in front of the radical, becuase x is positive.

For $\frac{\rho_t}{r} = 3$

$$x = 0.5\left(3 - 2y + \sqrt{9 + 12\frac{y}{\alpha^2}C_D}\right) \qquad (6_1)$$

Again, the values of x are found for different combinations of x and y and the values of $\frac{x}{y}$ plotted on Charts 26 and 28.

On chart 25, one case is illustrated for $\alpha = 60°$ and $y = 2 = \frac{S}{r}$. First, the slope $\varphi_{max} = 27°$ was selected; from the intersection of the vertical line from it with the curve for $\alpha = 60°$ a horizontal line shows that $\frac{R_c + r}{S} = 3.3$. If this line would be continued, it would intersect the vertical line from $y = 2$ well above the curve for $\alpha = 60°$ on the right side of the chart. It means the smallest radius ρ_c of the cam curvature is > 2, so the maximum slope φ_{max} determines the size of the cam.

Next, a reverse procedure is used. The vertical line from $y = 2$ is taken only to the intersection with the curve for $\alpha = 60°$. The intersection indicates the conditions when $\frac{\rho_t}{r} = 3$ or $\frac{\rho_c}{r} = 2$.

The horizontal line intersects the ordinates where $\frac{R_c + r}{S} = 2.35$, which will give the smallest value of R_c to have the radius of the curvature of the cam surface $\rho_c = 2$. If the horizontal line is continued, it intersects the line of the cam slope for $\alpha = 60°$ and this intersection shows that the angle φ_{max} of the slope will be 35°.

If the designer can increase the radius r of the cam follower, so $y = \frac{S}{r} = 1.5$, then for $\rho_c = 2$, the slope angle would be 29°. This example demonstrates the usefulness of charts of this type for a quick appraisal of possible changes in design.

When the value of $\frac{x}{y} = \frac{R_c + r}{S}$, is established, the minimum radius R_c of cam can be determined.

$$\frac{x}{y} = \frac{R_c + r}{S}; \qquad R_c = \frac{x}{y}\,.\,S - r = r\left(\frac{x}{y}y - 1\right) \qquad (7)$$

r and $y = \frac{S}{r}$ are known, while $\frac{x}{y}$ is selected from the chart.

Chart 25. Maximum Slope and Curvature

Modified Trapezoid; Diagram I, b = .25; P = 1.5

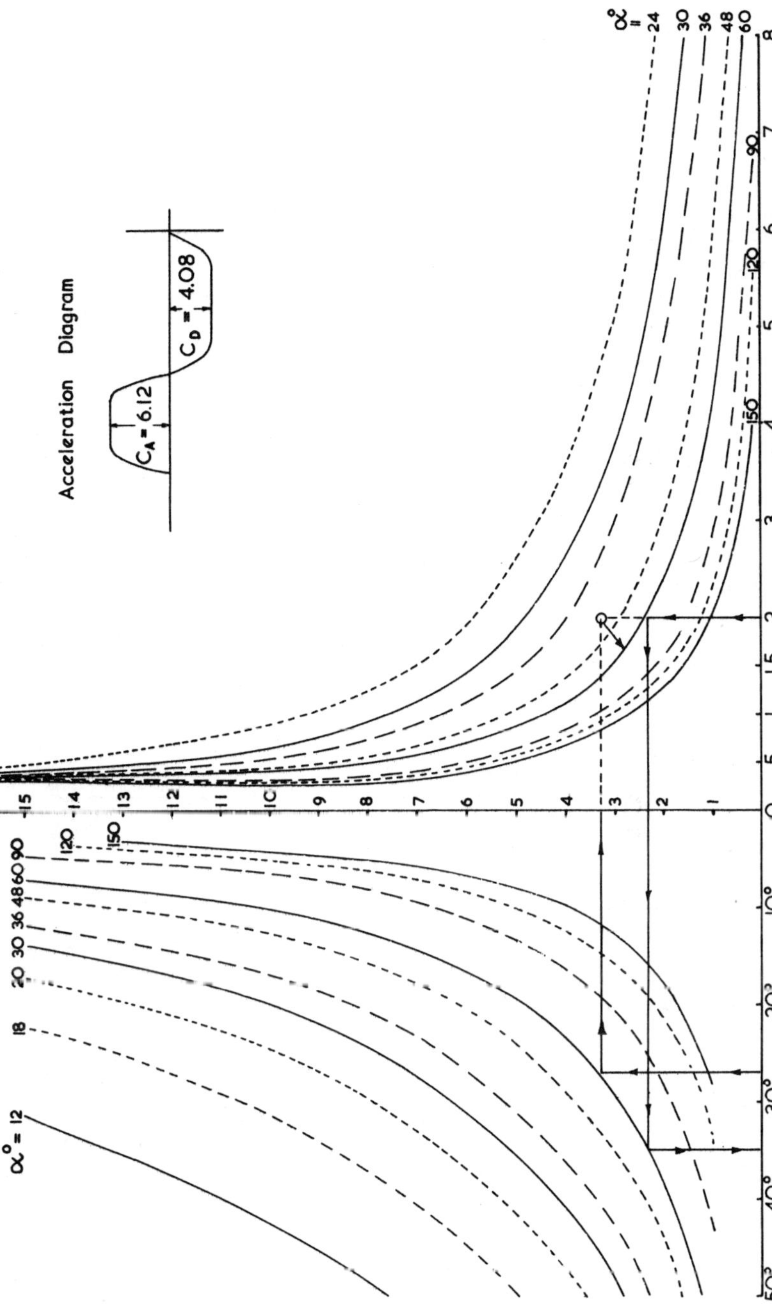

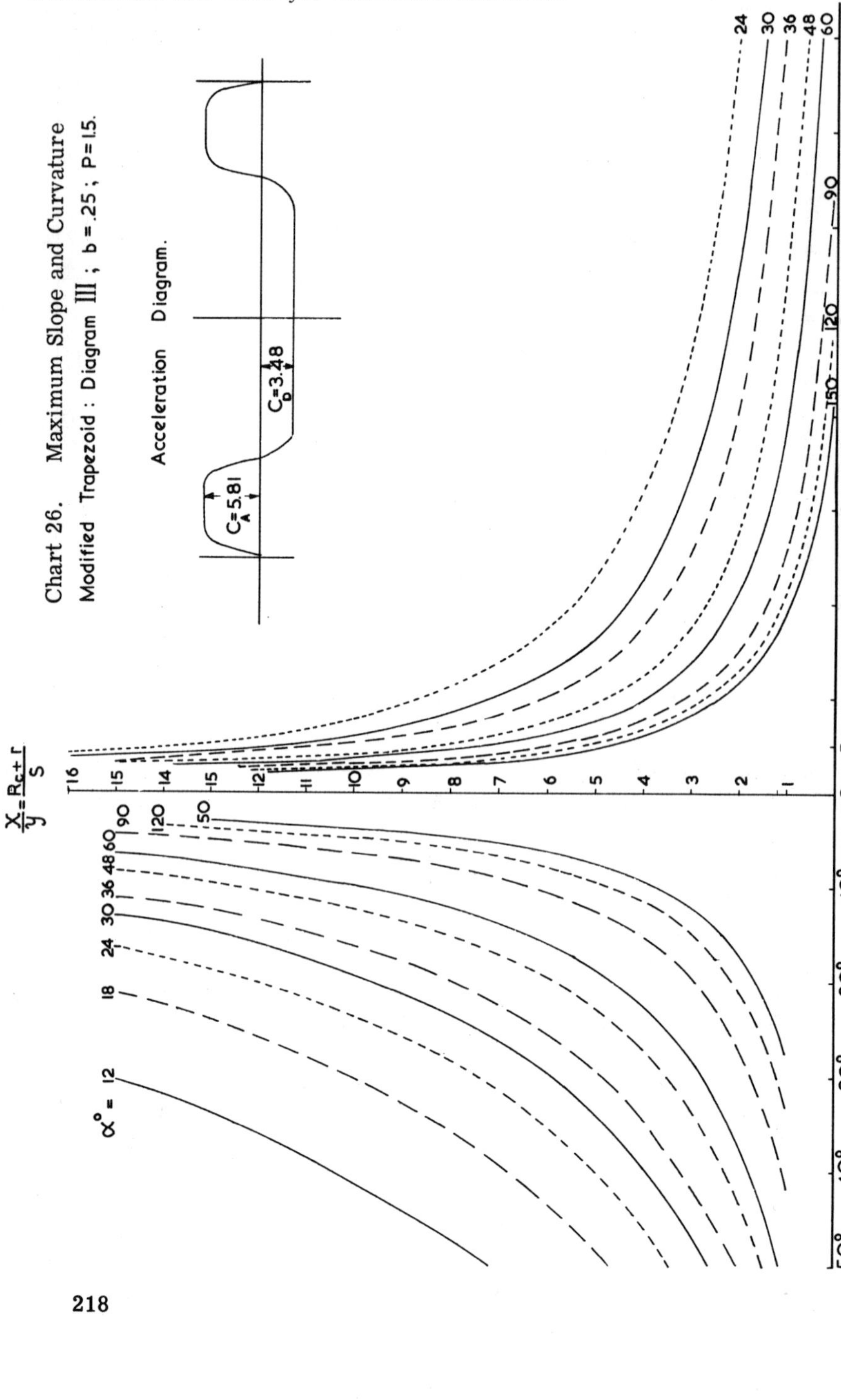

Chart 26. Maximum Slope and Curvature

Modified Trapezoid : Diagram III ; b = .25 ; P = 1.5.

Chart 27. Maximum Slope and Curvature

Modified Sine; Diagram I; c = .25; P = 1.0.

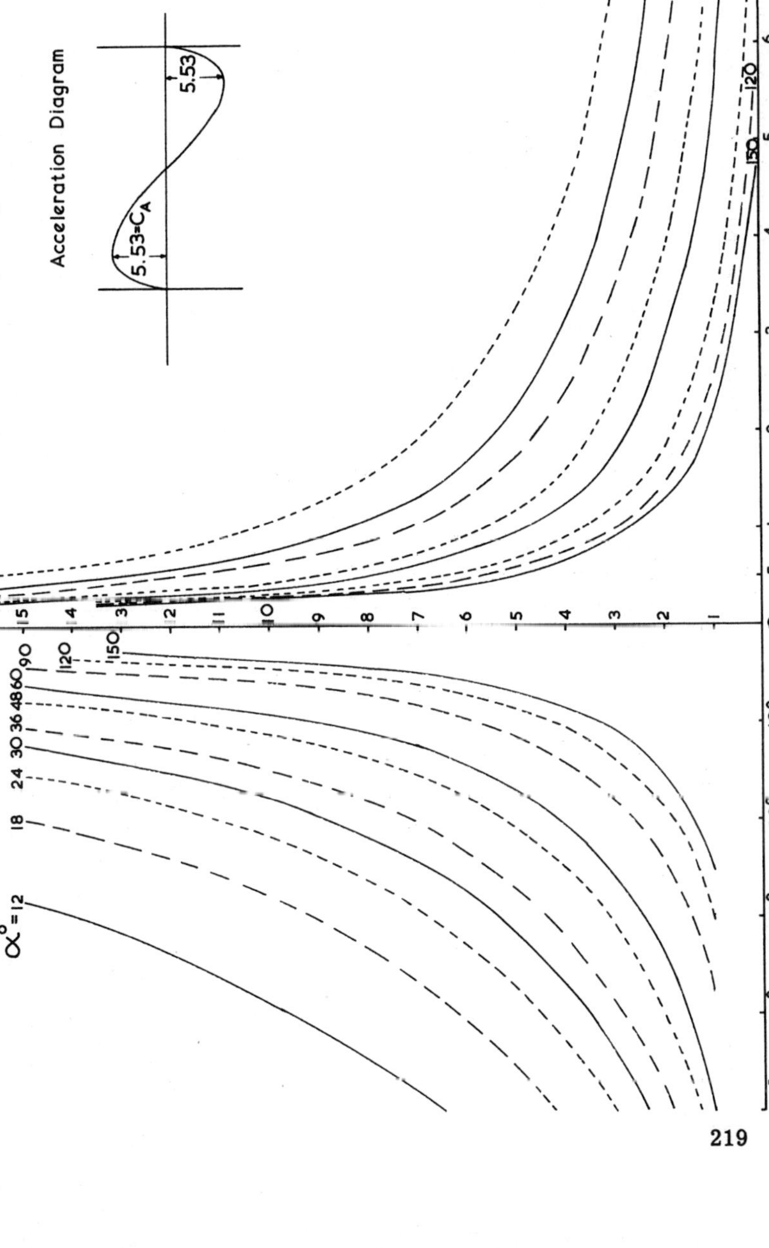

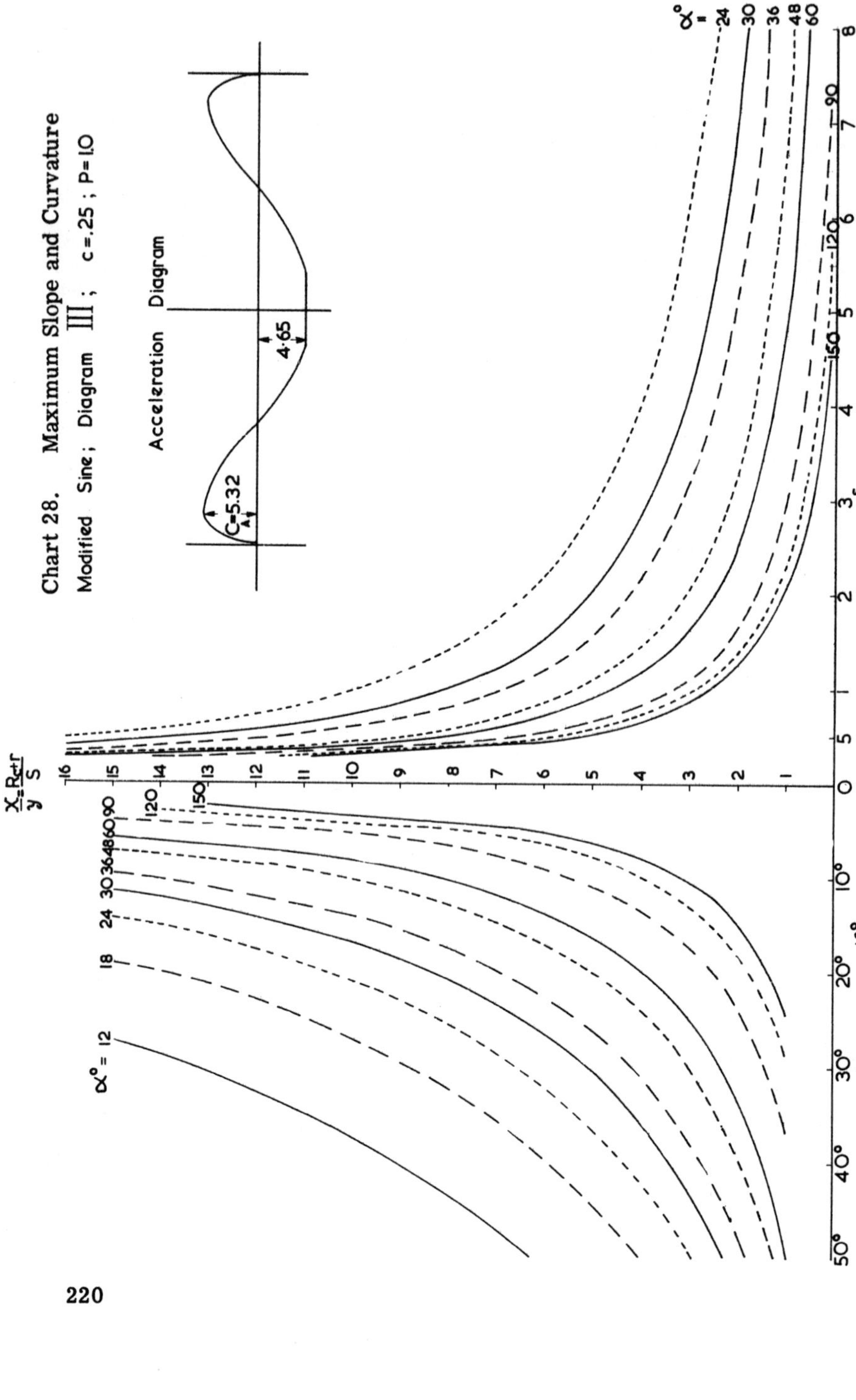

Chart 28. Maximum Slope and Curvature

Modified Sine; Diagram III; c = .25 ; P = 1.0

Calculation of the Torque on the Cam Shaft

When a cam is installed on a shaft, transmitting the power to other cams and mechanisms, quite often the torque created by the cam can be overlooked, if it represents only a few percent of the total load of the shaft.

When only one or a few heavily loaded cams are installed on the shaft, the knowledge of their torques becomes important. First—flat cams will be discussed as presenting more difficulties. The same nomenclature is used as in the analysis of the slope of the cam (page 206), because the torque is affected by the slope of the cam surface, as it was mentioned before. The torque creating force is

$$C = U \,.\, \tan\varphi$$

This force is applied on a radius $R_1 = R_c + r + s_1$, which connects centers of the cam and of the follower. Then the general expression for the torque on the cam shaft is

$$T_c = C \,.\, R_1 = U \,.\, \tan\varphi \left[R_c + r + S \,.\, f_s\left(\frac{t}{T}\right)\right] = U \,.\, S \,.\, \tan\varphi \,.$$

$$.\left[\frac{R_c + r}{S} + f_s\left(\frac{t}{T}\right)\right].$$

Introduce the equation (1) for $\tan\varphi$ (page 208)

$$T_c = U \,.\, S\left[\frac{R_c + r}{S} + f_s\left(\frac{t}{T}\right)\right] . \frac{180}{\alpha^\circ \pi} . \frac{C_v \,.\, f_s\left(\frac{t}{T}\right)}{\frac{R_c + r}{S} + f_s\left(\frac{t}{T}\right)} = U \,.\, S\,\frac{180}{\alpha^\circ \pi} \,.$$

$$.\, C_v \,.\, f_v\left(\frac{t}{T}\right). \qquad (1)$$

The torque is proportional to the force U and the stroke S; it is inversely proportional to the working period α° and it is affected by the selection of the acceleration diagram due to the presence of the coefficient C_v of the maximum velocity.

There are many types of forces acting on the cam and combinations of them are numerous; they can be separated into three groups, each characterized by the variation of their magnitude.

A. Constant forces created by friction or similar causes, as by a preload of the spring holding the cam follower against the cam surface.

B. Forces which change proportionally to the displacement of the cam follower as springs or pneumatic devices operated by the follower.

C. Dynamic forces affected by the following factors:

(a) characteristics of the selected acceleration diagram;

(b) mass or moment of inertia of the cam follower system;

(c) stroke and time during which it is performed.

These forces change between + and — maxima.

Forces, needed to do some work during the stroke, quite often are irregular so it is impossible to classify them.

It is possible to express the force U, as a group of components varying in different ways, but relations between their magnitudes may be different in every case. Therefore, it is more practical to write separately expressions for torques, created by three groups of forces.

A. Torque created by a constant force U_c.

From the equation (1) it is easy to write the maximum value of the torque because the only variable item—$f_v\left(\frac{t}{T}\right)$—changes between 0 and 1.

$$T_{cc\,max} = U_c \cdot S \cdot \frac{180}{\alpha^\circ \pi} \cdot C_v \tag{2}$$

The variation of this torque is shown on the Chart 29 for four "standard" acceleration diagrams in a dimensionless form.

$$\frac{T_{cc}}{U \cdot S \cdot \frac{180}{\alpha^\circ \pi}} = C_v \cdot f_v\left(\frac{t}{T}\right) \tag{2_1}$$

To use this chart, multiply the ordinates by $US\frac{180}{\alpha^\circ \pi} = U \cdot S\frac{1}{\alpha}$ with α in radians. The chart represents "Constant Force Cam Torque Factor."

B. Torque created by a force proportional to the displacement

The total force produced by the spring is

$$\text{Total force} = q + K \cdot b \cdot f_s\left(\frac{t}{T}\right)$$

q —is the preload of the spring;

K —is the spring factor in lb. per inch of deflection;

b —actual deflection of the spring when the cam follower moves on the stroke S.

$b \cdot f_s\left(\frac{t}{T}\right)$—represents the variation of the length of the spring during the stroke.

The variable part of the spring force is $U_s = K \cdot b \cdot f_s\left(\frac{t}{T}\right)$.

Then the torque is

$$T_{cs} = K \cdot b \cdot S \cdot \frac{180}{\alpha^\circ \pi} \cdot f_s\left(\frac{t}{T}\right) \cdot C_v \cdot f_v\left(\frac{t}{T}\right) \qquad (3)$$

Though both variable items—$f_s\left(\frac{t}{T}\right)$ and $f_v\left(\frac{t}{T}\right)$—vary between 0 and 1, their maxima do not coincide. The Chart 30 shows the variation of the torque for four acceleration diagrams in the dimensionless form.

$$\frac{T_{cs}}{KbS\,\frac{180}{\alpha\pi}} = f_s\left(\frac{t}{T}\right) \cdot C_v \cdot f_v\left(\frac{t}{T}\right) \qquad (3_1)$$

which is called "Displacement Force Cam Torque Factor."

The torque created by the preload of the spring q can be taken from the Chart 29 and in the equation (2_1) U_c is replaced by q.

C. Torque created by dynamic forces

The dynamic forces are caused by necessity to move masses lineally or rotationally. A cam follower may have a lineal or angular movement and it may produce a simple motion of a single mass or move different masses with strokes—lineal or angular—different than the stroke of the cam follower itself. All dynamic forces in the cam follower system should be replaced by one equivalent force acting on the center of the cam follower; in other words, all moving masses should be replaced by one equivalent mass moving with the stroke S of the cam follower.

Suppose that the equivalent force acting at the center of the follower expresses as the weight W in lbs; then the equivalent mass connected with the follower will be $\frac{W}{g}$, where g is the gravity acceleration = 386 inches—per sec^2.

The dynamic force acting on the cam follower will be

$$U_D = \frac{W}{g} \,.\, \text{Acceleration} = \frac{W}{g} \,.\, S\left(\frac{360}{\alpha^\circ} N\right)^2 . \; C_A \,.\, f_A\left(\frac{t}{T}\right).$$

Introduce U_D into the general equation (1) for the cam torque.

$$T_{CD} = \frac{W}{g} S^2\left(\frac{360}{\alpha^\circ} N\right)^2 . \; \frac{180}{\alpha^\circ \pi} \,.\, C_V \,.\, f_V\left(\frac{t}{T}\right) . \; C_A \,.\, f_A\left(\frac{t}{T}\right) \qquad (4)$$

The Chart 31 shows the dimensionless data

$$\frac{T_{CD}}{\frac{W}{g} \,.\, S^2 \,.\left(\frac{360}{\alpha^\circ} N\right)^2 \frac{180}{\alpha^\circ \pi}} = C_V \,.\, f_V\left(\frac{t}{T}\right) . \; C_A \,.\, f_A\left(\frac{t}{T}\right) \qquad (4_1)$$

which can be called "Dynamic Force Cam Torque Factor."

From the equation (2) it is possible to distinguish the part of the torque created by the slope of the cam $S \frac{180}{\alpha^\circ \pi} \,.\, C_V \,.\, f_V\left(\frac{t}{T}\right)$ and the part due to the dynamic force $\frac{W}{g} \,.\, S\left(\frac{360}{\alpha^\circ} N\right)^2 . \; C_A \,.\, f_A\left(\frac{t}{T}\right)$.

Torque on barrel cams

The force which creates the torque on the cam is the same as in the case of the flat cams

$$C = U \,.\, \tan \varphi$$

where the angle of the slope is simpler than for flat cams, equation (3) (page 209).

The force C is on a constant radius R, then the torque is

$$T_c = CR = UR\,\frac{S}{R}\cdot\frac{180}{\alpha^\circ\pi}\cdot C_V\cdot f_V\left(\frac{t}{T}\right) = US\,\frac{180}{\alpha^\circ\pi}\,C_V\cdot f_V\left(\frac{t}{T}\right)$$

The expression is exactly the same as for flat cams, equation (1) on page 221. Therefore, the charts for different cam torque factors can be used for barrel cams.

Different charts prepared for four acceleration diagrams, where Modified Trapezoid Diagram I and III have P = 1.5 and Modified Sine Diagrams I and III have P = 1. These charts do not permit a good comparison between the two groups of diagrams. Therefore, a Table is provided for comparison of torques created by dynamic forces, which are produced by different Acceleration Diagrams I.

	Sine Curve b = c = 0.5	Modified Sine c = 0.25	Modified Trapezoid b = 0.25
Comparative values of max. torques	1.0	0.76	1.27
Location of + and — maximum torques in terms of t/T	0.333 and 0.667	0.287 and 0.713	0.393 and 0.607
Coefficient of max. acceleration, C_A	6.283	5.528	4.888
Coefficient of max. velocity; C_V	2.0	1.76	2.0

It is evident that Modified Sine Diagram I is more advantageous when dynamic forces are predominant and have a high value:

(a) the maximum value of the torque on the cam is lower and

(b) the change from + max. to — max. value is slower.

Modified Trapezoid has the advantage of a lower dynamic force acting on the cam follower system and the cam torque, caused by a spring, is slightly lower than in case of a Modified Sine.

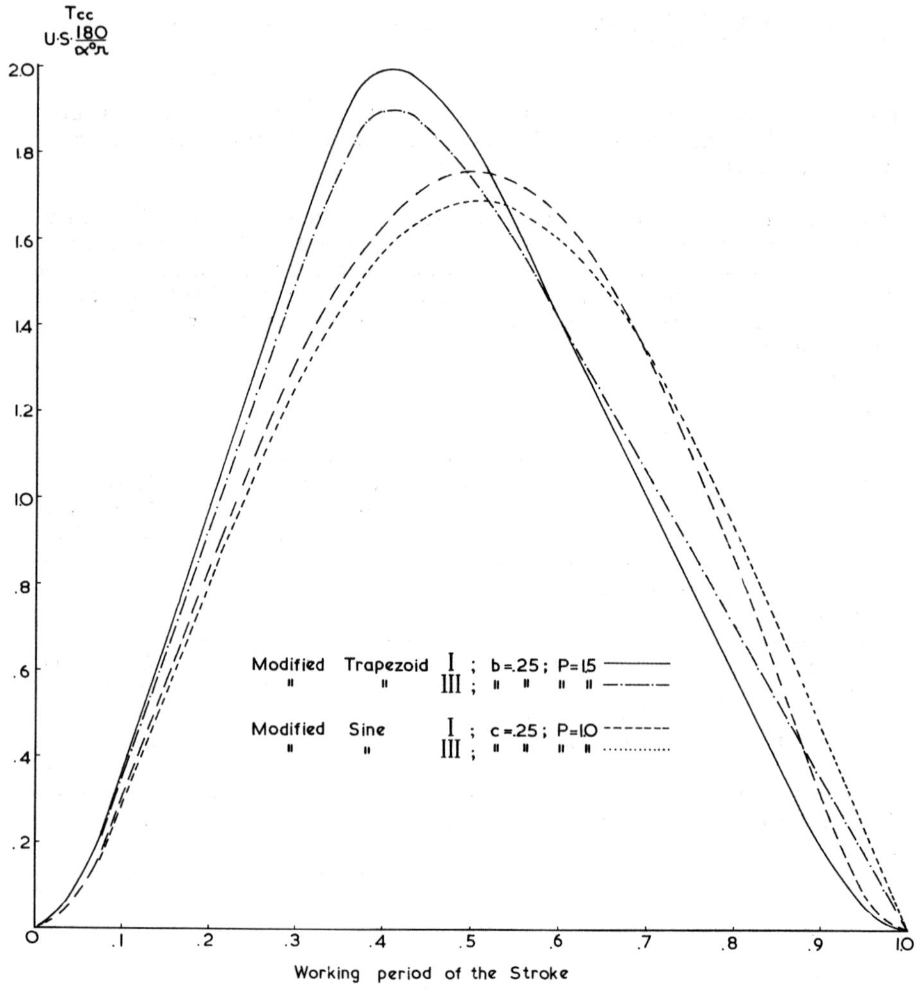

Chart 29. Constant Force Cam Torque Factor

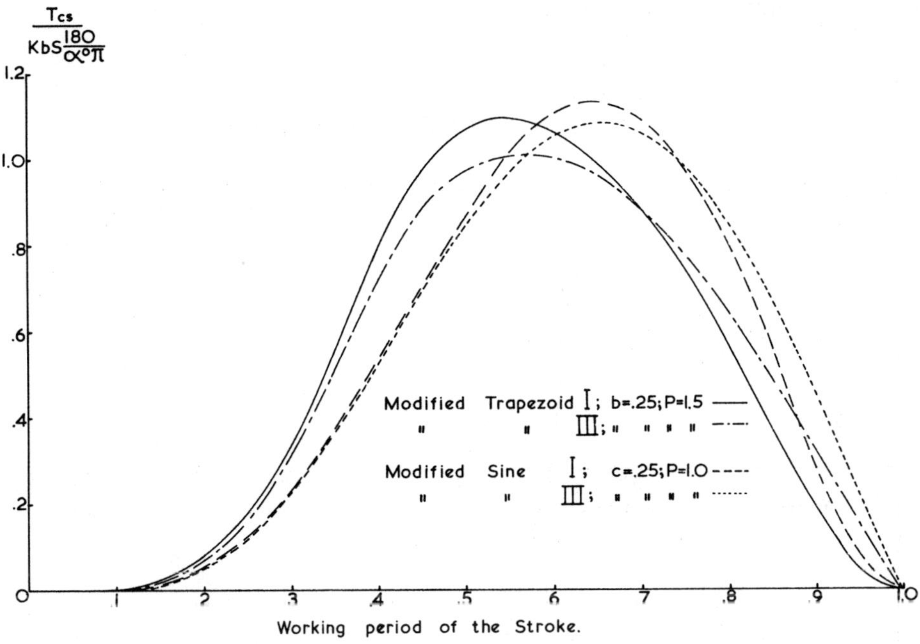

Chart 30. Displacement Force Cam Torque Factor

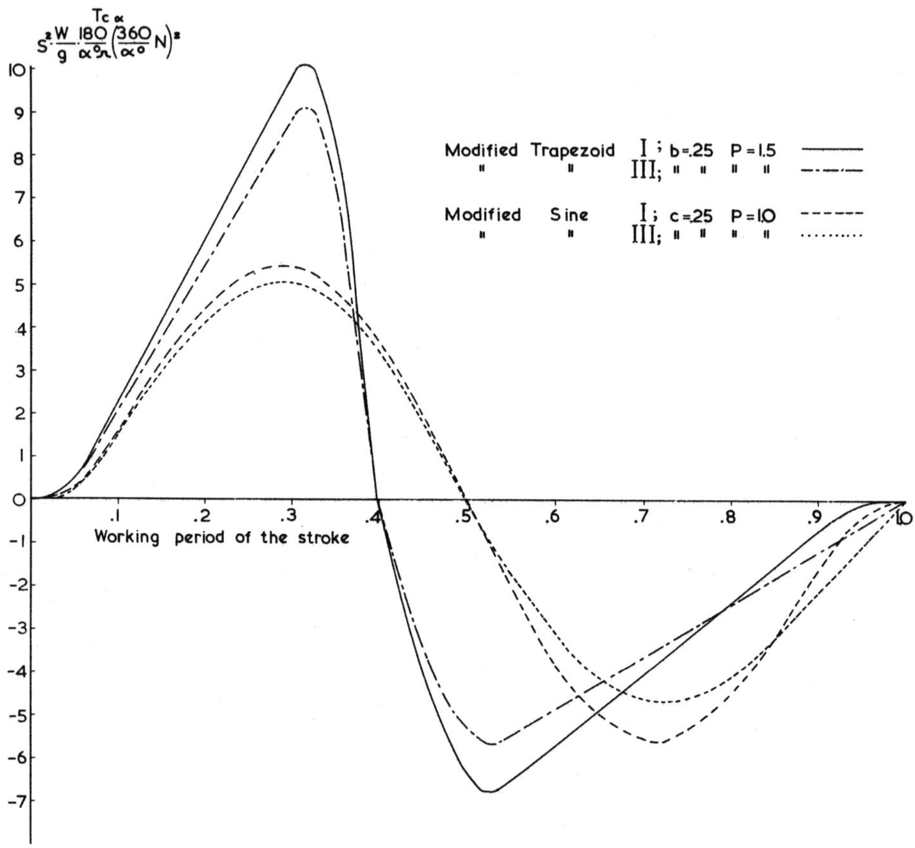

Chart 31. Dynamic Force Cam Torque Factor

Influence of Uniform Velocity Period on Torque Created by Dynamic Forces

In the Chapter III, Example X, the analysis was made of movements which have uniform velocity period occupying $x \cdot \alpha°$, where $\alpha°$ is the working period of the whole stroke S.

It was established that coefficients C_V and C_A should be changed, if equations for the velocity and for the acceleration retain the stroke S and the working period $\alpha°$.

$$C_{VX} = \frac{C_V}{1 + x\,(C_V - 1)};$$

and

$$C_{AX} = \frac{C_A}{[1 + x(C_V - 1)]\cdot(1 - x)}$$

These coefficients should be introduced into the equation (4) for the cam torque

$$T_{CD} = \underbrace{\left[\frac{W}{g}\cdot S\cdot\left(\frac{360}{\alpha°}N\right)^2 C_A\cdot f_A\left(\frac{t}{T}\right)\right]}_{\text{Factor created by dynamic forces}}\cdot\underbrace{\left[S\cdot\frac{180}{\alpha°\pi}\cdot C_V\cdot f_V\left(\frac{t}{T}\right)\right]}_{\text{Factor created by the slope of the cam}}\cdot$$

Factor created by dynamic forces

Factor created by the slope of the cam

$$\cdot\underbrace{\left[\frac{1}{[1 + x(C_V - 1)]^2\cdot(1 - x)}\right]}_{\text{Correction factor created by the presence of uniform velocity}}$$

Correction factor created by the presence of uniform velocity

The Table shows the influence of the length of the uniform velocity period on tan φ and T_{CD}; each item is shown for two diagrams: Modified Sine with $C_V = 1.76$ and Modified Trapezoid with $C_V = 2$.

$C_V =$	x =	0	0.1	0.2	0.3	1/3	0.4	0.5	0.6	0.8
2	tan φ for $x > 0$	1.0	0.(90)	0.8(3)	0.7692	0.75	0.7143	0.(6)	0.625	0.(5)
1.76	tan φ for $x = 0$	1.0	0.9294	0.868	0.8143	0.7979	0.7669	0.7246	0.6868	0.6215
2	T_{CD} for $x > 0$	1.0	0.9183	0.868	0.8453	0.844	0.85	0.(8)	0.9766	1.543
1.76	T_{CD} for $x = 0$	1.0	0.9597	0.9419	0.9473	0.9554	0.98	1.0502	1.18	1.934

Minimum value of the torque for Modified Trapezoid Acceleration is when $x = 1/3$ and for Modified Sine Acceleration when $x = 0.239$.

But the absolute value of T_{CD} for these points is lower for the Modified Sine, due to smaller slopes of the cam surface and a lower torque.

Conclusion

Only in rare cases it may be advisable to draw diagrams of combined torques, caused by three types of forces and by the work performed by the cam follower system. Usually, it is sufficient to determine maximum values of torques and combine them approximately; it is better to overestimate the resultant torque, than to underestimate it.

Greater strength and rigidity of the cam shaft will increase the frequency of free vibrations of the combination—shaft, cam, and the follower system—as a result there will be less distortion of the selected acceleration diagram due to vibrations of the cam shaft-cam system.

A reminder—all equations and numerical data are based on the system—

force (and weight) — in lb;

lineal dimensions — in inches;

time — in sec.

as a result

gravity acceleration $g = 386$ inch/sec^2

SUBJECT INDEX